IS ANYONE LISTENING
Aliens and the Universe

Charles W. Zamzow Jr., JD, PhD

Copyright © 2018 Charles Zamzow

All rights reserved.

ISBN-13: 9781720056461

© 2018 Charles Zamzow. Printed and bound in the United States of America All Rights reserved. No part of this book may be reproduced or transmitted in any form or by any means, electronic or mechanical, including photocopying, recording, or by an information storage and retrieval system—except by a reviewer who may quote brief passages in a review to be printed in a magazine or newspaper—without permission in writing from the publisher, or information, please contact Jupiter Technical Solutions, PO Box 335, Ottawa, IL. 61350.

Although the author and publisher have made every effort to ensure the accuracy and completeness of information contained in this hook, we assume no responsibility for errors, inaccuracies, omissions, or any inconsistency herein. Any slights of people, places, or organizations are unintentional.

ATTENTION CORPORATIONS, UNIVERSITIES, COLLEGES, AND PROFESSIONAL ORGANIZATIONS;

Quantity discounts are available on bulk purchases of this book for educational purposes Special books or book excerpts can also be created to the specific needs. For information, please contact Jupiter Technical Solutions, PO Box 335, Ottawa, IL. 61350

CONTENTS

	Introduction	10
1	Chapter	12
	The Drake Equation	
2	Chapter	23
	Extraterrestrial life	
3	Chapter	44
	The Fermi Paradox: Why Haven't We Found Alien Life Anywhere Else in the Universe?	
4	Chapter	51
	Search for Extraterrestrial Intelligence	
5	Chapter	76
	How Do Scientists Search for Extraterrestrial Life?	
6	Chapter	90
	Using Artificial Intelligence to Search for Extraterrestrial Intelligence	
7	Chapter	114
	Wernher Von Braun	
8	Chapter	120

Stephen Hawkins

9   Chapter   141
    Carl Sagon

    Conclusion NASA   149

    BIBLIOGRAPHY   167

    INDEX   201

ACKNOWLEDGMENTS

An author does not and should not write in a vacuum. Any book worth doing starts with the author's vision and commitment. It's then seasoned with the experience, advice, and inspiration of others. Some people actively participate in shaping the manuscript while others nurture the author's environment. In either case, no author can create a book without their help.

Special thanks to my wife Vivian for always believing in me.

To Vivian, my best friend and forever love

INTRODUCTION

A group of top NASAs scientists told the space agency's chief yesterday that the forthcoming generation of space telescopes are likely to discover habitable Earth-like worlds and probably alien life – perhaps within 20 years.

We're all someone's Sun

"Sometime in the near future, people will be able to point to a star and say, 'that star has a planet like Earth'," said Sara Seager, professor of planetary science and physics at MIT.

"Just imagine the moment, when we find potential signatures of life," concurred Matt Mountain, a top space telescope boffin.

"Imagine the moment when the world wakes up and the human race realizes that its long loneliness in time and space may be over - the possibility we're no longer alone in the universe."

"I think in the next 20 years we will find out we are not alone in the universe," added NASA astronomer Kevin Hand.

The assembled boffin were excited about the near future, at the prospect of the James Webb Space Telescope's deployment to the Earth-Sun L2 point, far beyond the Moon's orbit, where – among other things – it will be able to probe the atmospheres of far-flung exoplanets circling other suns. The mighty JWST, a collaboration between NASA and international allies, is currently planned to lift off in 2018.

Other exciting developments in the field of exoplanetology are to include the proposed Wide Field Infrared Survey Telescope - Astrophysics Focused Telescope Assets (WFIRST-AFTA) early in the next decade, and the Transiting Exoplanet Surveying Satellite (TESS) in 2017.

These various space observatories will to a large extent be carrying out in-depth investigation of exoplanets previously discovered by other means such as the Kepler spacecraft.

"This technology we are using to explore exoplanets is real," enthused John Grunsfeld, a physics PhD and veteran astronaut with five Shuttle flights logged, now employed as a NASA science chief. "The James Webb Space Telescope and the next advances are happening now. These are not dreaming - this is what we do at NASA."

The head of NASA himself, Charles Bolden (a former shuttle pilot and US Marine general) ranged himself firmly alongside his top boffins in the aliens-are-surely-out-there camp.

"It's highly improbable in the limitless vastness of the universe that we humans stand alone," he stated.

Chapter 1

The Drake Equation

"What do we need to know about to discover life in space?"
How can we estimate the number of technological civilizations that might exist among the stars? While working as a radio astronomer at the National Radio Astronomy Observatory in Green Bank, West Virginia, Dr. Frank Drake conceived an approach to bound the terms involved in estimating the number of technological civilizations that may exist in our galaxy. The Drake Equation, as it has become known, was first presented by Drake in 1961 and identifies specific factors thought to play a role in the development of such civilizations. Although there is no unique solution to this equation, it is a generally accepted tool used by the scientific community to examine these factors.
-- Frank Drake, 1961

Where,
N = The number of civilizations in the Milky Way Galaxy whose electromagnetic emissions are detectable.
R^* = The rate of formation of stars suitable for the development of intelligent life.
f_p = The fraction of those stars with planetary systems.

n_e = The number of planets, per solar system, with an environment suitable for life.
f_l = The fraction of suitable planets on which life appears.
f_i = The fraction of life bearing planets on which intelligent life emerges.
f_c = The fraction of civilizations that develop a technology that releases detectable signs of their existence into space.
L = The length of time such civilizations release detectable signals into space.

Within the limits of our existing technology, any practical search for distant intelligent life must necessarily be a search for some manifestation of a distant technology. In each of its last four decadal reviews, the National Research Council has emphasized the relevance and importance of searching for evidence of the electromagnetic signature of distant civilizations.

Besides illuminating the factors involved in such a search, the Drake Equation is a simple, effective tool for stimulating intellectual curiosity about the universe around us, for helping us to understand that life as we know it is the product of a natural, cosmic evolution, and for making us realize how much we are a part of that universe. A key goal of the SETI Institute is to further high-quality research that will yield additional information related to any of the factors of this fascinating equation.

History

In September 1959, physicists Giuseppe Cocconi and Philip Morrison published an article in the journal *Nature* with the provocative title "Searching for Interstellar Communications." Cocconi and Morrison argued that radio telescopes had become sensitive enough to pick up transmissions that might be broadcast into space by civilizations orbiting other stars. Such messages, they suggested, might be transmitted at a wavelength of 21 cm (1,420.4 MHz). This is the wavelength of radio emission by neutral hydrogen, the most common element in the universe, and they reasoned that other intelligences might see this as a logical landmark in the radio spectrum.

Two months later, Harvard University astronomy professor Harlow Shapley speculated on the number of inhabited planets in the universe, saying "The universe has 10 million, million, million suns (10 followed by 18 zeros) similar to our own. One in a million has planets around it. Only one in a million has the right combination of chemicals, temperature, water, days and nights to support planetary life as we know it. This calculation arrives at the estimated figure of 100 million worlds where life has been forged by evolution."

Seven months after Cocconi and Morrison published their article, Drake made the first systematic search for signals from extraterrestrial intelligent beings. Using the 25 m dish of the National Radio Astronomy Observatory in Green Bank, West Virginia, Drake monitored two nearby Sun-like stars: Epsilon Eridani and Tau Ceti. In this project, which he called Project Ozma, he slowly scanned frequencies close to the 21 cm wavelength for six hours per day from April to July 1960. The project was well designed, inexpensive, and simple by today's standards. It was also unsuccessful.

Soon thereafter, Drake hosted a "search for extraterrestrial intelligence" meeting on detecting their radio signals. The meeting was held at the Green Bank facility in 1961. The equation that bears Drake's name arose out of his preparations for the meeting.

As I planned the meeting, I realized a few day[s] ahead of time we needed an agenda. And so, I wrote down all the things you needed to know to predict how hard it's going to be to detect extraterrestrial life. And looking at them it became evident that if you multiplied all these together, you got a number, N, which is the number of detectable civilizations in our galaxy. This was aimed at the radio search, and not to search for primordial or primitive life forms. —Frank Drake.

The ten attendees were conference organizer J. Peter Pearman, Frank Drake, Philip Morrison, businessman and radio amateur Dana Atchley, chemist Melvin Calvin, astronomer Su-Shu Huang, neuroscientist John C. Lilly, inventor Barney Oliver, astronomer Carl Sagan and radio-astronomer Otto Struve. These participants dubbed themselves "**The Order of the Dolphin**" (because of Lilly's work on dolphin communication), and commemorated their first meeting with a plaque at the observatory hall.

The Drake equation amounts to a summary of the factors affecting the likelihood that we might detect radio-communication from intelligent extraterrestrial life. The last four parameters, f_l, f_i, f_c, and L, are not known and are very difficult to estimate, with values ranging over many orders of magnitude (see criticism). Therefore, the usefulness of the Drake equation is not in the solving, but rather in the contemplation of all the various concepts which scientists must incorporate when considering the question of life elsewhere, and gives the question of life elsewhere a basis for scientific analysis. The Drake equation is a statement that stimulates intellectual curiosity about the universe around us, for helping us to understand that life as we know it is the end product of a natural, cosmic evolution, and for helping us realize how much we are a part of that universe. What the equation and the search for life has done is focus science on some of the other questions about life in the universe, specifically abiogenesis, the development of multi-cellular life and the development of intelligence itself.

Within the limits of our existing technology, any practical search for distant intelligent life must necessarily be a search for some manifestation of a distant technology. After about 50 years, the Drake equation is still of seminal importance because it is a 'road map' of what we need to learn to solve this fundamental existential question. It also formed the backbone of astrobiology as a science; although speculation is entertained to give context, astrobiology concerns itself primarily with hypotheses that fit firmly into existing scientific theories. Some 50 years of SETI have failed to find anything, even though radio telescopes, receiver techniques, and computational abilities have improved enormously since the early 1960s, but it has been discovered, at least, that our galaxy is not teeming with very powerful alien transmitters continuously broadcasting near the 21 cm hydrogen frequency. No one could say this in 1961.

Current Estimates

This section discusses and attempts to list the best current estimates for the parameters of the Drake equation.

Rate of star creation in our galaxy, R_*

Latest calculations from NASA and the European Space Agency indicate that the current rate of star formation in our galaxy is about 0.68–1.45 $M_☉$ of material per year. To get the number of stars per year, this must account for the initial mass function (IMF) for stars, where the average new star mass is about 0.5 $M_☉$. This gives a star formation rate of about 1.5–3 stars per year.

Fraction of those stars that have planets, f_p [edit]

Recent analysis of microlensing surveys has found that f_p may approach 1—that is, stars are orbited by planets as a rule, rather than the exception; and that there are one or more bound planets per Milky Way star.

Average number of planets per star having planets that might support life, n_e [edit]

In November 2013, astronomers reported, based on *Kepler* space mission data, that there could be as many as 40 billion Earth-sized planets orbiting in the habitable zones of sun-like stars and red dwarf stars within the Milky Way Galaxy. 11 billion of these estimated planets may be orbiting sun-like stars. Since there are about 100 billion stars in the galaxy, this implies $f_p \cdot n_e$ is roughly 0.4. The nearest planet in the habitable zone may be as close as 12 light-years away, according to the scientists.

The consensus at the Green Bank meeting was that n_e had a minimum value between 3 and 5. Dutch astronomer Govert

Schilling has opined that this is optimistic. Even if planets are in the habitable zone, the number of planets with the right proportion of elements is difficult to estimate. Brad Gibson, Yeshe Fenner, and Charley Lineweaver determined that about 10% of star systems in the Milky Way galaxy are hospitable to life, by having heavy elements, being far from supernovae and being stable for a sufficient time.

The discovery of numerous gas giants in close orbit with their stars has introduced doubt that life-supporting planets commonly survive the formation of their stellar systems. So-called hot Jupiters may migrate from distant orbits to near orbits, in the process disrupting the orbits of habitable planets.

On the other hand, the variety of star systems that might have habitable zones is not just limited to solar-type stars and Earth-sized planets. It is now estimated that even tidally locked planets close to red dwarf stars might have habitable zones, although the flaring behavior of these stars might argue against this. The possibility of life on moons of gas giants (such as Jupiter's moon Europa, or Saturn's moon Titan) adds further uncertainty to this figure.

The authors of the rare Earth hypothesis propose a number of additional constraints on habitability for planets, including being in galactic zones with suitably low radiation, high star metallicity, and low enough density to avoid excessive asteroid bombardment. They also propose that it is necessary to have a planetary system with large gas giants which provide bombardment protection without a hot Jupiter; and a planet with plate tectonics, a large moon that creates tidal pools, and moderate axial tilt to generate seasonal variation.

Fraction of the above that go on to develop life, f_l

Geological evidence from the Earth suggests that f_l may be high; life on Earth appears to have begun around the same time as favorable conditions arose, suggesting that abiogenesis may be relatively common once conditions are right. However, this evidence only looks at the Earth (a single model planet), and contains anthropic bias, as the planet of study was not chosen randomly, but by the living organisms that already inhabit it (ourselves). From a classical hypothesis testing standpoint, there are zero degrees of freedom, permitting no valid estimates to be made. If life were to be found on Mars that developed independently from life on Earth it would imply a value for f_l close to 1. While this would raise the degrees of freedom from zero to one, there would remain a great deal of uncertainty on any estimate due to the small sample size, and the chance they are not independent.

Countering this argument is that there is no evidence for abiogenesis occurring more than once on the Earth — that is, all terrestrial life stems from a common origin. If abiogenesis were more common it would be speculated to have occurred more than once on the Earth. Scientists have searched for this by looking for bacteria that are unrelated to other life on Earth, but none have been found yet. It is also possible that life arose more than once, but that other branches were out-competed, or died in mass extinctions, or were lost in other ways. Biochemists Francis Crick and Leslie Orgel laid special emphasis on this uncertainty: "At the moment we have no means at all of knowing" whether we are "likely to be alone in the galaxy (Universe)" or whether "the galaxy may be pullulating with life of many different forms." As an alternative to abiogenesis on Earth, they proposed the hypothesis of directed panspermia, which states that Earth life began with "microorganisms sent here deliberately by a technological society on another planet, by means of a special long-range unmanned spaceship".

Fraction of the above that develops intelligent life, f_i

This value remains particularly controversial. Those who favor a low value, such as the biologist Ernst Mayr, point out that of the billions of species that have existed on Earth, only one has become intelligent and from this, infer a tiny value for f_i. Likewise, the

Rare Earth hypothesis, notwithstanding their low value for n_e above, also think a low value for f_i dominates the analysis. Those who favor higher values note the generally increasing complexity of life over time, concluding that the appearance of intelligence is almost inevitable, implying an f_i approaching 1. Skeptics point out that the large spread of values in this factor and others make all estimates unreliable. (See Criticism).

In addition, while it appears that life developed soon after the formation of Earth, the Cambrian explosion, in which a large variety of multicellular life forms came into being, occurred a considerable amount of time after the formation of Earth, which suggests the possibility that special conditions were necessary. Some scenarios such as the snowball Earth or research into the extinction events have raised the possibility that life on Earth is relatively fragile. Research on any past life on Mars is relevant since a discovery that life did form on Mars but ceased to exist might raise our estimate of f_l but would indicate that in half the known cases, intelligent life did not develop.

Estimates of f_i have been affected by discoveries that the Solar System's orbit is circular in the galaxy, at such a distance that it remains out of the spiral arms for tens of millions of years (evading radiation from novae). Also, Earth's large moon may aid the evolution of life by stabilizing the planet's axis of rotation.

Fraction of the above revealing their existence via signal release into space, f_c

For deliberate communication, the one example we have (the Earth) does not do much explicit communication, though there are some efforts covering only a tiny fraction of the stars that might look for our presence. (See Arecibo message, for example). There is considerable speculation why an extraterrestrial civilization might exist but choose not to communicate. However, deliberate communication is not required, and calculations indicate that current or near-future Earth-level technology might well be detectable to civilizations not too much more advanced than our own. By this standard, the Earth is a communicating civilization.

Another question is what percentage of civilizations in the galaxy are close enough for us to detect, if they send out signals. For example, existing Earth radio telescopes could only detect Earth radio transmissions from roughly a light year away.

Lifetime of such a civilization wherein it communicates its signals into space, L

Michael Shermer estimated L as 420 years, based on the duration of sixty historical Earthly civilizations. Using 28 civilizations more recent than the Roman Empire, he calculates a figure of 304 years for "modern" civilizations. It could also be argued from Michael Shermer's results that the fall of most of these civilizations was followed by later civilizations that carried on the technologies, so it is doubtful that they are separate civilizations in the context of the Drake equation. In the expanded version, including *reappearance number*, this lack of specificity in defining single civilizations does not matter for the result, since such a civilization turnover could be described as an increase in the *reappearance number* rather than increase in L, stating that a civilization reappears in the form of the succeeding cultures. Furthermore, since none could communicate over interstellar space, the method of comparing with historical civilizations could be regarded as invalid.

David Grinspoon has argued that once a civilization has developed enough, it might overcome all threats to its survival. It will then last for an indefinite period, making the value for L potentially billions of years. If this is the case, then he proposes that the Milky Way galaxy may have been steadily accumulating advanced civilizations since it formed. He proposes that the last factor L be replaced with $f_{IC} \cdot T$, where f_{IC} is the fraction of communicating civilizations become "immortal" (in the sense that they simply do not die out), and T representing the length of time during which this process has been going on. This has the advantage that T

would be a relatively easy to discover number, as it would simply be some fraction of the age of the universe.

It has also been hypothesized that once a civilization has learned of a more advanced one, its longevity could increase because it can learn from the experiences of the other.

The astronomer Carl Sagan speculated that all of the terms, except for the lifetime of a civilization, are relatively high and the determining factor in whether there are large or small numbers of civilizations in the universe is the civilization lifetime, or in other words, the ability of technological civilizations to avoid self-destruction. In Sagan's case, the Drake equation was a strong motivating factor for his interest in environmental issues and his efforts to warn against the dangers of nuclear warfare.

Range of Results

As many skeptics have pointed out, the Drake equation can give a very wide range of values, depending on the assumptions, and the values used in portions of the Drake equation are not well-established. The result can be $N \ll 1$, meaning we are likely alone in the galaxy, or $N \gg 1$, implying there are many civilizations we might contact. One of the few points of wide agreement is that the presence of humanity implies a probability of intelligence arising of greater than zero.

As an example of a low estimate, combining NASA's star formation rates, the rare Earth hypothesis value of $f_p \cdot n_e \cdot f_l = 10^{-5}$, Mayr's view on intelligence arising, Drake's view of communication, and Shermer's estimate of lifetime:

$$R_* = 1.5\text{--}3 \text{ yr}^{-1},^{[26]} f_p \cdot n_e \cdot f_l = 10^{-5},^{[40]} f_i = 10^{-9},^{[43]} f_c = 0.2^{[\text{Drake, above}]}, \text{ and } L = 304 \text{ years}$$

gives:

$$N = 1.5 \times 10^{-5} \times 10^{-9} \times 0.2 \times 304 = 9.1 \times 10^{-11}$$

i.e., suggesting that we are probably alone in this galaxy, and possibly in the observable universe.

On the other hand, with larger values for each of the parameters above, values of N can be derived that are greater than 1. The following higher values that have been proposed for each of the parameters:

$R_* = 1.5\text{–}3 \text{ yr}^{-1}, f_p = 1, n_e = 0.2, f_l = 0.13, f_i = 1, f_c = 0.2$, and $L = 10^9$ years

Use of these parameters gives:

$N = 3 \times 1 \times 0.2 \times 0.13 \times 1 \times 0.2 \times 10^9 = 15{,}600{,}000$

Monte Carlo simulations of estimates of the Drake equation factors based on a stellar and planetary model of the Milky Way have resulted in the number of civilizations varying by a factor of 100.

Has Intelligent life *ever* Existed?

The Drake equation can be modified to determine just how unlikely intelligent life must be, to give the result that Earth has the only intelligent life that has ever arisen, either in our galaxy or the universe. This simplifies the calculation by removing the lifetime and communication constraints. Since star and planets counts are known, this leaves the only unknown as the odds that a habitable planet *ever* develops intelligent life. For Earth to have the only civilization that has ever occurred in the universe, then the odds of any habitable planet ever developing such a civilization must be less than 2.5×10^{-24}. Similarly, for Earth to host the only civilization in our galaxy for all time, the odds of a habitable zone planet ever hosting intelligent life must be less than 1.7×10^{-11} (about 1 in 60 billion). The figure for the universe implies that it is highly unlikely that Earth hosts the only intelligent life that has ever occurred. The figure for our galaxy suggests that other

Chapter 2

Extraterrestrial Life

Some major international efforts to search for extraterrestrial life.

- The search for extrasolar planets (image: *Kepler* telescope)
- Listening for extraterrestrial signals indicating intelligence (image: Allen array)
- Robotic exploration of the Solar System (image: *Curiosity* rover on Mars)

Extraterrestrial life, also called **alien life** (or, if it is a sentient or relatively complex individual, an "extraterrestrial" or "alien"), is life that occurs outside of Earth and that probably did not originate from Earth. These hypothetical life forms may range from simple prokaryotes to beings with civilizations far more advanced than humanity. The Drake equation speculates about the existence of intelligent life elsewhere in the universe. The science of extraterrestrial life in all its forms is known as exobiology.

Since the mid-20th century, there has been an ongoing search for signs of extraterrestrial life. This encompasses a search for current and historic extraterrestrial life, and a narrower search for extraterrestrial intelligent life. Depending on the category of search, methods range from the analysis of telescope and specimen data to radios used to detect and send communication signals.

The concept of extraterrestrial life, and particularly extraterrestrial intelligence, has had a major cultural impact, chiefly in works of science fiction. Over the years, science fiction communicated scientific ideas, imagined a wide range of possibilities, and influenced public interest in and perspectives of extraterrestrial life. One shared space is the debate over the wisdom of attempting communication with extraterrestrial intelligence. Some encourage aggressive methods to try for contact with intelligent extraterrestrial life. Others—citing the tendency of technologically advanced human societies to enslave or wipe out less advanced societies—argue that it may be dangerous to actively call attention to Earth.

Alien life, such as microorganisms, has been hypothesized to exist in the Solar System and throughout the universe. This hypothesis relies on the vast size and consistent physical laws of the observable universe. According to this argument, made by scientists such as Carl Sagan and Stephen Hawking, as well as well-regarded thinkers such as Winston Churchill, it would be improbable for life *not* to exist somewhere other than Earth. This argument is embodied in the Copernican principle, which states that Earth does not occupy a unique position in the Universe, and the mediocrity principle, which states that there is nothing special about life on Earth. The chemistry of life may have begun shortly after the Big Bang, 13.8 billion years ago, during a habitable epoch when the universe was only 10–17 million years old. Life may have emerged independently at many places throughout the universe. Alternatively, life may have formed less frequently, then spread—by meteoroids, for example—between habitable planets in a process called panspermia. In any case, complex organic molecules may have formed in the protoplanetary disk of dust grains surrounding the Sun before the formation of Earth. According to these studies, this process may occur outside Earth on several planets and moons of the Solar System and on planets of other stars.

Since the 1950s, scientists have proposed that "habitable zones" around stars are the most likely places to find life. Numerous discoveries in such zones since 2007 have generated numerical estimates of Earth-like planets —in terms of composition—of many billions. As of 2013, only a few planets have been discovered in these zones. Nonetheless, on 4 November 2013, astronomers reported, based on *Kepler* space mission data, that there could be as many as 40 billion Earth-sized planets orbiting in the habitable zones of Sun-like stars and red dwarfs in the Milky Way, 11 billion of which may be orbiting Sun-like stars. The nearest such planet may be 12 light-years away, according to the scientists. Astrobiologists have also considered a "follow the energy" view of potential habitats.

Evolution

A study published in 2017 suggests that due to how complexity evolved in species on Earth, the level of predictability for alien evolution elsewhere would make them look like life on our planet. One of the study authors, Sam Levin, notes "Like humans, we predict that they are made-up of a hierarchy of entities, which all cooperate to produce an alien. At each level of the organism there will be mechanisms in place to eliminate conflict, maintain cooperation, and keep the organism functioning. We can even offer some examples of what these mechanisms will be." There is also research in assessing the capacity of life for developing intelligence. It has been suggested that this capacity arises with the number of potential niches a planet contains, and that the complexity of life itself is reflected in the information density of planetary environments, which in turn can be computed from its niches.

Biochemical Basis

Life on Earth requires water as a solvent in which biochemical reactions take place. Sufficient quantities of carbon and other elements, along with water, might enable the formation of living organisms on terrestrial planets with a chemical make-up and temperature range similar to that of Earth. More generally, life based on ammonia (rather than water) has been suggested, though this solvent appears less suitable than water. It is also conceivable that there are forms of life whose solvent is a liquid hydrocarbon, such as methane, ethane or propane.

About 29 chemical elements play an active positive role in living organisms on Earth. About 95% of living matter is built upon only six elements: carbon, hydrogen, nitrogen, oxygen, phosphorus and sulfur. These six elements form the basic building blocks of virtually all life on Earth, whereas most of the remaining elements are found only in trace amounts. The unique characteristics of carbon make it unlikely that it could be replaced, even on another planet, to generate the biochemistry necessary for life. The carbon atom has the unique ability to make four strong chemical bonds with other atoms, including other carbon atoms. These covalent bonds have a direction in space, so that carbon atoms can form the skeletons of complex 3-dimensional structures with definite architectures such as nucleic acids and proteins. Carbon forms more compounds than all other elements combined. The great versatility of the carbon atom makes it the element most likely to provide the bases—even exotic ones—for the chemical composition of life on other planets.

Planetary Habitability in the Solar System

Some bodies in the Solar System have the potential for an environment in which extraterrestrial life can exist, particularly those with possible subsurface oceans. Should life be discovered elsewhere in the Solar System, astrobiologists suggest that it will more likely be in the form of extremophile microorganisms. According to NASA's 2015 Astrobiology Strategy, "Life on other worlds is most likely to include microbes, and any complex living system elsewhere is likely to have arisen from and be founded upon microbial life. Important insights on the limits of microbial life can be gleaned from studies of microbes on modern Earth, as well as their ubiquity and ancestral characteristics."

Mars may have niche subsurface environments where microbial life might exist. A subsurface marine environment on Jupiter's moon Europa might be the most likely habitat in the Solar System, outside Earth, for extremophile microorganisms.

The panspermia hypothesis proposes that life elsewhere in the Solar System may have a common origin. If extraterrestrial life was found on another body in the Solar System, it could have originated from Earth just as life on Earth could have been seeded from elsewhere (exogenesis). The first known mention of the term 'panspermia' was in the writings of the 5th century BC Greek philosopher Anaxagoras. In the 19th century it was again revived in modern form by several scientists, including Jöns Jacob Berzelius (1834), Kelvin (1871), Hermann von Helmholtz (1879) and, somewhat later, by Svante Arrhenius (1903). Sir Fred Hoyle (1915–2001) and Chandra Wickramasinghe (born 1939) are important proponents of the hypothesis who further contended that life forms continue to enter Earth's atmosphere, and may be responsible for epidemic outbreaks, new diseases, and the genetic novelty necessary for macroevolution.

Directed panspermia concerns the deliberate transport of microorganisms in space, sent to Earth to start life here, or sent from Earth to seed new stellar systems with life. The Nobel prize winner Francis Crick, along with Leslie Orgel proposed that seeds of life may have been purposely spread by an advanced extraterrestrial civilization, but considering an early "RNA world" Crick noted later that life may have originated on Earth.

Venus

In the early 20th century, Venus was often thought to be similar to Earth in terms of habitability, but observations since the beginning of the Space Age have revealed that Venus's surface is inhospitable to Earth-like life. However, between an altitude of 50 and 65 kilometers, the pressure and temperature are Earth-like, and it has been speculated that thermoacidophile extremophile microorganisms might exist in the acidic upper layers of the Venusian atmosphere. Furthermore, Venus likely had liquid water on its surface for at least a few million years after its formation.

Mars

Life on Mars has been long speculated. Liquid water is widely thought to have existed on Mars in the past, and now can occasionally be found as low-volume liquid brines in shallow Martian soil. The origin of the potential biosignature of methane observed in Mars' atmosphere is unexplained, although hypotheses not involving life have also been proposed.

There is evidence that Mars had a warmer and wetter past: dried-up river beds, polar ice caps, volcanoes, and minerals that form in the presence of water have all been found. Nevertheless, present conditions on Mars' subsurface may support life. Evidence obtained by the *Curiosity* rover studying Aeolis Palus, Gale Crater in 2013 strongly suggests an ancient freshwater lake that could have been a hospitable environment for microbial life.

Current studies on Mars by the *Curiosity* and *Opportunity* rovers are searching for evidence of ancient life, including a biosphere based on autotrophic, chemotrophic and/or chemolithoautotrophic microorganisms, as well as ancient water, including fluvio-lacustrine environments (plains related to ancient rivers or lakes) that may have been habitable. The search for evidence of habitability, taphonomy (related to fossils), and organic carbon on Mars is now a primary NASA objective.

Ceres

Ceres, the only dwarf planet in the asteroid belt, has a thin water-vapor atmosphere. Frost on the surface may also have been detected in the form of bright spots. The presence of water on Ceres has led to speculation that life may be possible there.

Jupiter System

Jupiter

Carl Sagan and others in the 1960s and 1970s computed conditions for hypothetical microorganisms living in the atmosphere of Jupiter. The intense radiation and other conditions, however, do not appear to permit encapsulation and molecular biochemistry, so life there is thought unlikely. In contrast, some of Jupiter's moons may have habitats capable of sustaining life. Scientists have indications that heated subsurface oceans of liquid water may exist deep under the crusts of the three outer Galilean moons—Europa, Ganymede, and Callisto. The EJSM/Laplace mission is planned to determine the habitability of these environments.

Europa

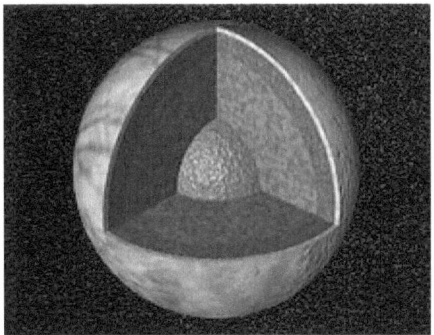

Internal structure of Europa. The blue is a subsurface ocean. Such subsurface oceans could possibly harbor life.

Jupiter's moon Europa has been subject to speculation about the existence of life due to the strong possibility of a liquid water ocean beneath its ice surface. Hydrothermal vents on the bottom of the ocean, if they exist, may warm the ice and could be capable of supporting multicellular microorganisms. It is also possible that Europa could support aerobic macrofauna using oxygen created by cosmic rays impacting its surface ice.

The case for life on Europa was greatly enhanced in 2011 when it was discovered that vast lakes exist within Europa's thick, icy shell. Scientists found that ice shelves surrounding the lakes appear to be collapsing into them, thereby providing a mechanism through which life-forming chemicals created in sunlit areas on Europa's surface could be transferred to its interior.

On 11 December 2013, NASA reported the detection of "clay-like minerals" (specifically, phyllosilicates), often associated with organic materials, on the icy crust of Europa. The presence of the minerals may have been the result of a collision with an asteroid or comet according to the scientists. The *Europa Clipper*, which would assess the habitability of Europa, is planned for launch in 2025. Europa's subsurface ocean is considered the best target for the discovery of life.

Saturn System

Titan and Enceladus have been speculated to have possible habitats supportive of life.

Enceladus

Enceladus, a moon of Saturn, has some of the conditions for life, including geothermal activity and water vapor, as well as possible under-ice oceans heated by tidal effects. The *Cassini–Huygens* probe detected carbon, hydrogen, nitrogen and oxygen—all key elements for supporting life—during its 2005 flyby through one of Enceladus's geysers spewing ice and gas. The temperature and density of the plumes indicate a warmer, watery source beneath the surface.

Titan

Titan, the largest moon of Saturn, is the only known moon in the Solar System with a significant atmosphere. Data from the *Cassini–Huygens* mission refuted the hypothesis of a global hydrocarbon ocean, but later demonstrated the existence of liquid hydrocarbon lakes in the polar regions—the first stable bodies of surface liquid discovered outside Earth. Analysis of data from the mission has uncovered aspects of atmospheric chemistry near the surface that are consistent with—but do not prove—the hypothesis that organisms there if present, could be consuming hydrogen, acetylene and ethane, and producing methane.

Small Solar System bodies

Small Solar System bodies have also been speculated to host habitats for extremophiles. Fred Hoyle and Chandra Wickramasinghe have proposed that microbial life might exist on comets and asteroids.

Other bodies

Models of heat retention and heating via radioactive decay in smaller icy Solar System bodies suggest that Rhea, Titania, Oberon, Triton, Pluto, Eris, Sedna, and Orcus may have oceans underneath solid icy crusts approximately 100 km thick. Of interest in these cases is the fact that the models indicate that the liquid layers are in direct contact with the rocky core, which allows efficient mixing of minerals and salts into the water. This is in contrast with the oceans that may be inside larger icy satellites like Ganymede, Callisto, or Titan, where layers of high-pressure phases of ice are thought to underlie the liquid water layer.

Hydrogen sulfide has been proposed as a hypothetical solvent for life and is quite plentiful on Jupiter's moon Io, and may be in liquid form a short distance below the surface.

Scientific search

The scientific search for extraterrestrial life is being carried out both directly and indirectly. As of September 2017, 3,667 exoplanets in 2,747 systems have been identified, and other planets and moons in our own solar system hold the potential for hosting primitive life such as microorganisms.

Direct search

Scientists search for biosignatures within the Solar System by studying planetary surfaces and examining meteorites. Some claim to have identified evidence that microbial life has existed on Mars. An experiment on the two Viking Mars landers reported gas emissions from heated Martian soil samples that some scientists argue are consistent with the presence of living microorganisms. Lack of corroborating evidence from other experiments on the same samples, suggests that a non-biological reaction is a more likely hypothesis. In 1996, a controversial report stated that structures resembling nanobacteria were discovered in a meteorite, ALH84001, formed of rock ejected from Mars.

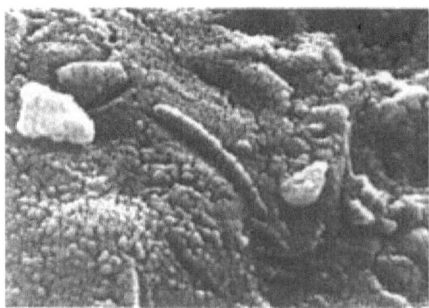

Electron micrograph of Martian meteorite ALH84001 showing structures that some scientists think could be fossilized bacteria-like life forms.

In February 2005, NASA scientists reported that they may have found some evidence of present life on Mars. The two scientists, Carol Stoker and Larry Lemke of NASA's Ames Research Center, based their claim on methane signatures found in Mars's atmosphere resembling the methane production of some forms of primitive life on Earth, as well as on their own study of primitive life near the Rio Tinto river in Spain. NASA officials soon distanced NASA from the scientists' claims, and Stoker herself backed off from her initial assertions. Though such methane findings are still debated, support among some scientists for the existence of life on Mars exists.

In November 2011, NASA launched the Mars Science Laboratory that landed the *Curiosity* rover on Mars. It is designed to assess the past and present habitability on Mars using a variety of scientific instruments. The rover landed on Mars at Gale Crater in August 2012.

The Gaia hypothesis stipulates that any planet with a robust population of life will have an atmosphere in chemical disequilibrium, which is relatively easy to determine from a distance by spectroscopy. However, significant advances in the ability to find and resolve light from smaller rocky worlds near their star are necessary before such spectroscopic methods can be used to analyze extrasolar planets. To that effect, the Carl Sagan Institute was founded in 2014 and is dedicated to the atmospheric characterization of exoplanets in circumstellar habitable zones. Planetary spectroscopic data will be obtained from telescopes like WFIRST and ELT.

In August 2011, findings by NASA, based on studies of meteorites found on Earth, suggest DNA and RNA components (adenine, guanine and related organic molecules), building blocks for life as we know it, may be formed extraterrestrials in outer space. In October 2011, scientists reported that cosmic dust contains complex organic matter ("amorphous organic solids with a mixed aromatic-aliphatic structure") that could be created naturally, and rapidly, by stars. One of the scientists suggested that these compounds may have been related to the development of life on Earth and said that, "If this is the case, life on Earth may have had an easier time getting started as these organics can serve as basic ingredients for life."

In August 2012, and in a world first, astronomers at Copenhagen University reported the detection of a specific sugar molecule, glycolaldehyde, in a distant star system. The molecule was found around the protostellar binary *IRAS 16293-2422*, which is located 400 light years from Earth. Glycolaldehyde is needed to form ribonucleic acid, or RNA, which is similar in function to DNA. This finding suggests that complex organic molecules may form in stellar systems prior to the formation of planets, eventually arriving on young planets early in their formation.

Indirect search

Projects such as SETI are monitoring the galaxy for electromagnetic interstellar communications from civilizations on other worlds. If there is an advanced extraterrestrial civilization, there is no guarantee that it is transmitting radio communications in the direction of Earth or that this information could be interpreted as such by humans. The length of time required for a signal to travel across the vastness of space means that any signal detected would come from the distant past.

The presence of heavy elements in a star's light-spectrum is another potential biosignature; such elements would (in theory) be found if the star was being used as an incinerator/repository for nuclear waste products.

Extrasolar Planets

Artist's Impression of Gliese 581 c, the first terrestrial extrasolar planet discovered within its star's habitable zone.

Artist's impression of the Kepler telescope in space.

Some astronomers search for extrasolar planets that may be conducive to life, narrowing the search to terrestrial planets within the habitable zone of their star. Since 1992 over two thousand exoplanets have been discovered (3,758 planets in 2,808 planetary systems including 627 multiple planetary systems as of 1 April 2018). The extrasolar planets so far discovered range in size from that of terrestrial planets similar to Earth's size to that of gas giants larger than Jupiter. The number of observed exoplanets is expected to increase greatly in the coming years.

The *Kepler* space telescope has also detected a few thousand candidate planets, of which about 11% may be false positives.

There is at least one planet on average per star. About 1 in 5 Sun-like stars have an "Earth-sized" planet in the habitable zone, with the nearest expected to be within 12 light-years distance from Earth. Assuming 200 billion stars in the Milky Way, that would be 11 billion potentially habitable Earth-sized planets in the Milky Way, rising to 40 billion if red dwarfs are included. The rogue planets in the Milky Way possibly number in the trillions.

The nearest known exoplanet is Proxima Centauri b, located 4.2 light-years (1.3 pc) from Earth in the southern constellation of Centaurus.

As of March 2014, the least massive planet known is PSR B1257+12 A, which is about twice the mass of the Moon. The most massive planet listed on the NASA Exoplanet Archive is DENIS-P J082303.1-491201 b, about 29 times the mass of Jupiter, although according to most definitions of a planet, it is too massive to be a planet and may be a brown dwarf instead. Almost all of the planets detected so far are within the Milky Way, but there have also been a few possible detections of extragalactic planets. The study of planetary habitability also considers a wide range of other factors in determining the suitability of a planet for hosting life.

One sign that a planet probably already contains life is the presence of an atmosphere with significant amounts of oxygen, since that gas is highly reactive and generally would not last long without constant replenishment. This replenishment occurs on Earth through photosynthetic organisms. One way to analyze the atmosphere of an exoplanet is through spectrograph when it transits its star, though this might only be feasible with dim stars like white dwarfs.

Terrestrial Analysis

The science of astrobiology considers life on Earth as well, and in the broader astronomical context. In 2015, "remains of biotic life" were found in 4.1 billion-year-old rocks in Western Australia, when the young Earth was about 400 million years old. According to one of the researchers, "If life arose relatively quickly on Earth, then it could be common in the universe."

Cultural Impact

Cosmic Pluralism

Cosmic pluralism, the plurality of worlds, or simply pluralism, describes the philosophical belief in numerous "worlds" in addition to Earth, which might harbor extraterrestrial life. Before the development of the heliocentric theory and a recognition that the Sun is just one of many stars, the notion of pluralism was largely mythological and philosophical. Medieval Muslim writers like Fakhr al-Din al-Razi and Muhammad al-Baqir supported cosmic pluralism on the basis of the Qur'an.

With the scientific and Copernican revolutions, and later, during the Enlightenment, cosmic pluralism became a mainstream notion, supported by the likes of Bernard le Bovier de Fontenelle in his 1686 work *Entretiens sur la pluralité des mondes*. Pluralism was also championed by philosophers such as John Locke, Giordano Bruno and astronomers such as William Herschel. The astronomer Camille Flammarion promoted the notion of cosmic pluralism in his 1862 book *La pluralité des mondes habités*. None of these notions of pluralism were based on any specific observation or scientific information.

Early Modern Period

There was a dramatic shift in thinking initiated by the invention of the telescope and the Copernican assault on geocentric cosmology. Once it became clear that Earth was merely one planet amongst countless bodies in the universe, the theory of extraterrestrial life started to become a topic in the scientific community. The best known early-modern proponent of such ideas was the Italian philosopher Giordano Bruno, who argued in the 16th century for an infinite universe in which every star is surrounded by its own planetary system. Bruno wrote that other worlds "have no less virtue nor a nature different to that of our earth" and, like Earth, "contain animals and inhabitants".

In the early 17th century, the Czech astronomer Anton Maria Schyrleus of Rheita mused that "if Jupiter has (...) inhabitants (...) they must be larger and more beautiful than the inhabitants of Earth, in proportion to the [characteristics] of the two spheres".

In Baroque literature such as *The Other World: The Societies and Governments of the Moon* by Cyrano de Bergerac, extraterrestrial societies are presented as humoristic or ironic parodies of earthly society. The didactic poet Henry More took up the classical theme of the Greek Democritus in "Democritus Platonissans, or an Essay Upon the Infinity of Worlds" (1647). In "The Creation: a Philosophical Poem in Seven Books" (1712), Sir Richard Blackmore observed: "We may pronounce each orb sustains a race / Of living things adapted to the place". With the new relative viewpoint that the Copernican revolution had wrought, he suggested "our world's sunne / Becomes a starre elsewhere". Fontanelle's "Conversations on the Plurality of Worlds" (translated into English in 1686) offered similar excursions on the possibility of extraterrestrial life, expanding, rather than denying, the creative sphere of a Maker.

The possibility of extraterrestrials remained a widespread speculation as scientific discovery accelerated. William Herschel, the discoverer of Uranus, was one of many 18th–19th-century astronomers who believed that the Solar System is populated by alien life. Other luminaries of the period who championed "cosmic pluralism" included Immanuel Kant and Benjamin Franklin. At the height of the Enlightenment, even the Sun and Moon were considered candidates for extraterrestrial inhabitants.

19th century

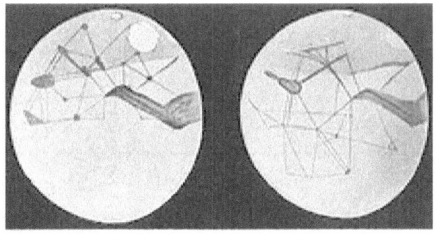

Artificial Martian channels, depicted by Percival Lowell

Speculation about life on Mars increased in the late 19th century, following telescopic observation of apparent Martian canals—which soon, however, turned out to be optical illusions. Despite this, in 1895, American astronomer Percival Lowell published his book *Mars,* followed by *Mars and its Canals* in 1906, proposing that the canals were the work of a long-gone civilization. The idea of life on Mars led British writer H. G. Wells to write the novel *The War of the Worlds* in 1897, telling of an invasion by aliens from Mars who were fleeing the planet's desiccation.

Spectroscopic analysis of Mars's atmosphere began in earnest in 1894, when U.S. astronomer William Wallace Campbell showed that neither water nor oxygen was present in the Martian atmosphere. By 1909 better telescopes and the best perihelic opposition of Mars since 1877 conclusively put an end to the canal hypothesis.

The science fiction genre, although not so named during the time, developed during the late 19th century. Jules Verne's *Around the Moon* (1870) features a discussion of the possibility of life on the Moon, but with the conclusion that it is barren. Stories involving extraterrestrials are found in e.g. Garrett P. Serviss's *Edison's Conquest of Mars* (1898), an unauthorized sequel to *The War of the Worlds* by H. G. Wells was published in 1897 which stands at the beginning of the popular idea of the "Martian invasion" of Earth prominent in 20th-century pop culture.

20th Century

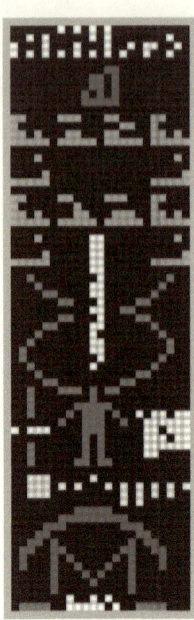

The Arecibo message is a digital message sent to Messier 13, and is a well-known symbol of human attempts to contact extraterrestrials.

Most unidentified flying objects or UFO sightings can be readily explained as sightings of Earth-based aircraft, known astronomical objects, or as hoaxes. Nonetheless, a certain fraction of the public believe that UFOs might be of extraterrestrial origin, and, indeed, the notion has had influence on popular culture.

The possibility of extraterrestrial life on the Moon was ruled out in the 1960s, and during the 1970s it became clear that most of the other bodies of the Solar System do not harbor highly developed life, although the question of primitive life on bodies in the Solar System remains open.

Recent History

The failure so far of the SETI program to detect an intelligent radio signal after decades of effort has at least partially dimmed the prevailing optimism of the beginning of the space age. Notwithstanding, belief in extraterrestrial beings continues to be voiced in pseudoscience, conspiracy theories, and in popular folklore, notably "Area 51" and legends. It has become a pop culture trope given less-than-serious treatment in popular entertainment.

In the words of SETI's Frank Drake, "All we know for sure is that the sky is not littered with powerful microwave transmitters".[180] Drake noted that it is entirely possible that advanced technology results in communication being carried out in some way other than conventional radio transmission. At the same time, the data returned by space probes, and giant strides in detection methods, have allowed science to begin delineating habitability criteria on other worlds, and to confirm that at least other planets are plentiful, though aliens remain a question mark. The Wow! signal, detected in 1977 by a SETI project, remains a subject of speculative debate.

In 2000, geologist and paleontologist Peter Ward and astrobiologist Donald Brownlee published a book entitled *Rare Earth: Why Complex Life is Uncommon in the Universe*. In it, they discussed the Rare Earth hypothesis, in which they claim that Earth-like life is rare in the universe, whereas microbial life is common. Ward and Brownlee are open to the idea of evolution on other planets that is not based on essential Earth-like characteristics (such as DNA and carbon).

Theoretical physicist Stephen Hawking in 2010 warned that humans should not try to contact alien life forms. He warned that aliens might pillage Earth for resources. "If aliens visit us, the outcome would be much as when Columbus landed in America, which didn't turn out well for the Native Americans", he said. Jared Diamond had earlier expressed similar concerns.

In November 2011, the White House released an official response to two petitions asking the U.S. government to acknowledge formally that aliens have visited Earth and to disclose any intentional withholding of government interactions with extraterrestrial beings. According to the response, "The U.S. government has no evidence that any life exists outside our planet, or that an extraterrestrial presence has contacted or engaged any member of the human race." Also, according to the response, there is "no credible information to suggest that any evidence is being hidden from the public's eye." The response noted "odds are pretty high" that there may be life on other planets but "the odds of us making contact with any of them—especially any intelligent ones—are extremely small, given the distances involved."

In 2013, the exoplanet Kepler-62f was discovered, along with Kepler-62e and Kepler-62c. A related special issue of the journal *Science*, published earlier, described the discovery of the exoplanets.

On 17 April 2014, the discovery of the Earth-size exoplanet Kepler-186f, 500 light-years from Earth, was publicly announced; it is the first Earth-size planet to be discovered in the habitable zone and it has been hypothesized that there may be liquid water on its surface.

On 13 February 2015, scientists (including Geoffrey Marcy, Seth Shostak, Frank Drake and David Brin) at a convention of the American Association for the Advancement of Science, discussed Active SETI and whether transmitting a message to possible intelligent extraterrestrials in the Cosmos was a good idea; one result was a statement, signed by many, that a "worldwide scientific, political and humanitarian discussion must occur before any message is sent".

On 20 July 2015, British physicist Stephen Hawking and Russian billionaire Yuri Milner, along with the SETI Institute, announced a well-funded effort, called the Breakthrough Initiatives, to expand efforts to search for extraterrestrial life. The group contracted the services of the 100-meter Robert C. Byrd Green Bank Telescope in

West Virginia in the United States and the 64-meter Parkes Telescope in New South Wales, Australia.

Chapter 3

The Fermi Paradox: Why Haven't We Found Alien Life Anywhere Else in the Universe?

In the Milky Way alone, there are an estimated 100,000 million stars and 100 billion planets, many of which are "potentially habitable," meaning they could have the conditions right for life to emerge.

With an estimated two trillion galaxies in the universe, the universe should be teeming with alien life. So where is it?

It is important to remember when scientists discuss alien life, there are two types—microbial and intelligent. When NASA talks about searching for life elsewhere in our solar system, such as on Saturn's icy moon Enceladus, they are referring to microbial life—simple primitive cells or multicellular organisms—not intelligent beings that have developed something resembling what we find on Earth.

Microbial life can exist in some of the most extreme conditions on Earth—from the bottom of the ocean to boiling hot hydrothermal vents. If we apply our understanding of life on Earth to conditions that may exist on other planets, the likelihood of life being able to survive increases.

Then there's intelligent life. Intelligence does not necessarily mean human intelligence. It could come in a variety of forms. Octopus, for example, are extremely intelligent by scientists. But we couldn't have a conversation with an octopus as it possesses a different form of intelligence to our own.

The Milky Way seen from the White Desert outside Cairo. Amr Dalsh/Retuers

There is also an argument to say that we, as a species, are not very intelligent. On the Kardashev scale—a measure of a civilization's level of intelligence based on how technologically advanced it is—mankind is not even really at level one. The most advanced civilization would have mastered interstellar travel and can harness and control the energy produced by its entire galaxy.

The universe has existed for an estimated 13.82 billion years, with the first stars forming around 500 million years later. After the stars came the planets, with Earth being a late addition, having been created 4.6 billion years ago.

With countless potentially habitable planets forming over the history of the universe, the chance life emerged on just one of them—Earth—is highly unlikely. But if extraterrestrial life is everywhere, why do we seem to be alone? This conundrum is known as the Fermi Paradox.

Dhara Patel, Astronomer for Royal Observatory Greenwich, tells *Newsweek*: " *Our* nearest star is a star called Proxima Centauri and last year in August they found an exoplanet going around this star. It's 4.2 light years away. If we're talking about current technology, the technology we have with our spacecrafts, it would take 70,000 years to get there.

The planet Proxima b orbits the red dwarf star Proxima Centauri, the closest star to our Solar System, as depicted in this artist's impression released by the European Southern Observatory on August 24, 2016. New study indicates the planet may well have a climate right for alien life. ESO/M. Kornmesser/Handout

"Life on Earth only started around four billion years ago. Our universe is 13.8 billion years old and the first life could've started about one to two billion years after that start of the universe, so there's been a huge time before the Earth where civilizations could've developed, could've been wiped out and we wouldn't have even known. Are we in the right time? Have we missed a civilization in the past or is there a civilization yet to come that we could communicate with?"

In the meantime, researchers are working to identify planets that sit within the habitable zone of their solar systems. This is the region of space where liquid water could exist, as it is not too hot or too cold.

NASA's James Webb Telescope, which is scheduled to launch next year, will be able to detect the atmospheres of these planets, giving a stronger indication of whether they could host life. The space agency is also planning a mission to Jupiter's moon Europa to search out microbial life—which could be the first proof that life exists elsewhere in our solar system.

But until our technology develops to the point where we can explore space on far shorter timescales, the chance of finding intelligent life is very much in the alien ball court.

In 1950, a learned lunchtime conversation set the stage for decades of astronomical exploration. Physicist Enrico Fermi submitted to his colleagues around the table a couple contentions, summarized as 1) The galaxy is very old and very large, with hundreds of billions of stars and likely even more habitable planets. 2) That means there should be more than enough time for advanced civilizations to develop and flourish across the galaxy.

So, where the heck are they?

This simple, yet powerful argument became known as the Fermi Paradox, and it still boggles many sage minds today. Aliens should be common, yet there is no convincing evidence that they exist.

Here are twelve possible reasons why this is so.

1. There aren't any aliens to find. As unlikely as it seems in a galaxy with hundreds of billions of stars and as many as 40 billion Earth-size planets in habitable zones, we could be alone.

2. There is no intelligent life besides us. (This assumes, of course, that humans count as intelligent.) Life may exist, but it could simply take the form of miniscule microbes or other cosmically "quiet" animals.

3. Intelligent species lack advanced technology. Currently, astronomers utilize radio telescopes to listen intently to the night sky. So, if alien species aren't broadcasting any signals, we'd never know they existed.

4. Intelligent life self-destructs. Whether via weapons of mass destruction, planetary pollution, or manufactured virulent disease, it may be the nature of intelligent species to commit suicide, existing for only a short time before winking out of existence.

5. The universe is a deadly place. On cosmic timescales – think billions of years – life may be fleeting. All it takes is a single asteroid, supernova, gamma ray burst, or solar flare to render a life-harboring planet lifeless.

6. Space is big. The Milky Way alone is 100,000 light years across, so it's conceivable that the focused signals of intelligent aliens, which are limited to the speed of light, simply haven't reached us yet.

7. We haven't been looking long enough. Eighty years. That's the amount of time that radio telescopes, which allow us to detect alien signals, have been around. And we've been actively searching for aliens for maybe sixty years. That's not very long at all.

8. We're not looking in the correct place. As previously mentioned, space is big, so there are tons of regions to listen for alien signals. If we're not listening precisely in the direction from which a signal is originating, we'd never hear it. As Andrew Fain explained at *Universe Today*, it's like trying to speak with your friend on a 250,000,000,000-channel CB radio, without any knowledge of the frequency on which they are transmitting. You'll probably be channel flipping for a long time.

9. Alien technology may be too advanced. Radio technology may be commonplace here on Earth, but on far-flung worlds, alien societies may have graduated to more advanced communication technologies, like neutrino signals. We can't decipher those just yet.

10. Nobody is transmitting. Instead, everybody may be listening. That's basically how it is here on Earth. Apart from a few paltry efforts to broadcast strong signals over a narrow frequency band towards the stars above, we've barely made our presence known in the universe. In fact, if aliens have radio telescopes like what we have on Earth, our television and radio broadcasts would only be detectable up to 0.3 light-years away. That distance doesn't even transcend the farthest reaches of our solar system.

11. Earth is deliberately not being contacted. On Earth, we have policies about contacting indigenous peoples; it's possible that the same thing could be happening with us. Just like in Star Trek, advanced alien societies may enforce rules that limit contact only to species that attain a lofty degree of technological or cultural evolution.

12. Aliens are already here, and we just don't realize it. Conspiracy theorists love this unlikely explanation. While the chances are remote, it's not *impossible* that government agencies are concealing the presence of aliens. Although it's more likely that aliens are already amongst us, observing humanity in the clever and ironic guise of lab mice.

Chapter 4

Search for Extraterrestrial Intelligence

The **search for extraterrestrial intelligence** (**SETI**) is a collective term for scientific searches for intelligent extraterrestrial life, for example, monitoring electromagnetic radiation for signs of transmissions from civilizations on other planets.

Scientific investigation began shortly after the advent of radio in the early 1900s, and focused international efforts have been going on since the 1980s. In 2015 Stephen Hawking and Russian billionaire Yuri Milner announced a well-funded effort called the Breakthrough Initiatives.

History

Early Work

There have been many earlier searches for extraterrestrial intelligence within the Solar System. In 1896, Nikola Tesla suggested that an extreme version of his wireless electrical transmission system could be used to contact beings on Mars. In 1899, while conducting experiments at his Colorado Springs experimental station, he thought he had detected a signal from that planet since an odd repetitive static signal seemed to cut off when Mars set in the night sky. Analysis of Tesla's research has ranged from suggestions that Tesla detected nothing, he simply misunderstood the new technology he was working with, to claims that Tesla may have been observing signals from Marconi's European radio experiments and even that he could have picked up naturally occurring Jovian plasma torus signals. In the early 1900s,

Guglielmo Marconi, Lord Kelvin and David Peck Todd also stated their belief that radio could be used to contact Martians, with Marconi stating that his stations had also picked up potential Martian signals.

On August 21–23, 1924, Mars entered an opposition closer to Earth than at any time in the century before or the next 80 years. In the United States, a "National Radio Silence Day" was promoted during a 36-hour period from August 21–23, with all radios quiet for five minutes on the hour, every hour. At the United States Naval Observatory, a radio receiver was lifted 3 kilometers (1.9 miles) above the ground in a dirigible tuned to a wavelength between 8 and 9 km, using a "radio-camera" developed by Amherst College and Charles Francis Jenkins. The program was led by David Peck Todd with the military assistance of Admiral Edward W. Eberle (Chief of Naval Operations), with William F. Friedman (chief cryptographer of the United States Army), assigned to translate any potential Martian messages.

A 1959 paper by Philip Morrison and Giuseppe Cocconi first pointed out the possibility of searching the microwave spectrum, and proposed frequencies and a set of initial targets.

In 1960, Cornell University astronomer Frank Drake performed the first modern SETI experiment, named "Project Ozma", after the Queen of Oz in L. Frank Baum's fantasy books. Drake used a radio telescope 26 meters (85 ft) in diameter at Green Bank, West Virginia, to examine the stars Tau Ceti and Epsilon Eridani near the 1.420 gigahertz marker frequency, a region of the radio spectrum dubbed the "water hole" due to its proximity to the hydrogen and hydroxyl radical spectral lines. A 400 kilohertz band around the marker frequency was scanned, using a single-channel receiver with a bandwidth of 100 hertz. He found nothing of interest.

Soviet scientists took a strong interest in SETI during the 1960s and performed a number of searches with omnidirectional antennas in the hope of picking up powerful radio signals. Soviet astronomer Iosif Shklovsky wrote the pioneering book in the field, *Universe, Life, Intelligence* (1962), which was expanded upon by American astronomer Carl Sagan as the best-selling book *Intelligent Life in the Universe* (1966).

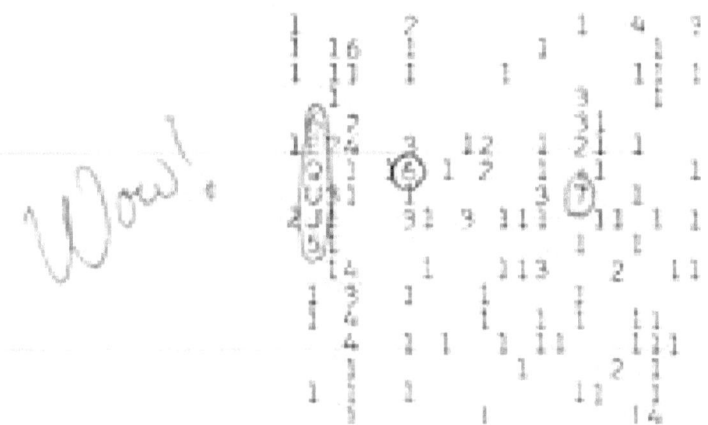

The WOW! Signal
Credit: The Ohio State University Radio Observatory and the North American AstroPhysical Observatory (NAAPO).

In the March 1955 issue of *Scientific American*, John D. Kraus described an idea to scan the cosmos for natural radio signals using a flat-plane radio telescope equipped with a parabolic reflector. Within two years, his concept was approved for construction by Ohio State University. With a total of US$71,000 in grants from the National Science Foundation, construction began on an 8-hectare (20-acre) plot in Delaware, Ohio. This Ohio State University Radio Observatory telescope was called "Big Ear". Later, it began the world's first continuous SETI program, called the Ohio State University SETI program.

In 1971, NASA funded a SETI study that involved Drake, Bernard M. Oliver of Hewlett-Packard Corporation, and others. The resulting report proposed the construction of an Earth-based radio telescope array with 1,500 dishes known as "Project Cyclops". The price tag for the Cyclops array was US$10 billion. Cyclops was not built, but the report formed the basis of much SETI work that followed.

The Ohio State SETI program gained fame on August 15, 1977, when Jerry Ehman, a project volunteer, witnessed a startlingly strong signal received by the telescope. He quickly circled the indication on a printout and scribbled the exclamation "Wow!" in the margin. Dubbed the *Wow! signal*, it is considered by some to be the best candidate for a radio signal from an artificial, extraterrestrial source ever discovered, but it has not been detected again in several additional searches.

Sentinel, META, and BETA

In 1980, Carl Sagan, Bruce Murray, and Louis Friedman founded the U.S. Planetary Society, partly as a vehicle for SETI studies.

In the early 1980s, Harvard University physicist Paul Horowitz took the next step and proposed the design of a spectrum analyzer specifically intended to search for SETI transmissions. Traditional desktop spectrum analyzers were of little use for this job, as they sampled frequencies using banks of analog filters and so were restricted in the number of channels they could acquire. However, modern integrated-circuit digital signal processing (DSP) technology could be used to build autocorrelation receivers to check far more channels. This work led in 1981 to a portable spectrum analyzer named "Suitcase SETI" that had a capacity of 131,000 narrow band channels. After field tests that lasted into 1982, Suitcase SETI was put into use in 1983 with the 26-meter (85 ft) Harvard/Smithsonian radio telescope at Oak Ridge Observatory in Harvard, Massachusetts. This project was named "Sentinel" and continued into 1985.

Even 131,000 channels were not enough to search the sky in detail at a fast rate, so Suitcase SETI was followed in 1985 by Project "META", for "Mega channel Extra-Terrestrial Assay". The META spectrum analyzer had a capacity of 8.4 million channels and a channel resolution of 0.05 hertz. An important feature of META was its use of frequency Doppler shift to distinguish between signals of terrestrial and extraterrestrial origin. The project was led by Horowitz with the help of the Planetary Society, and was partly funded by movie maker Steven Spielberg. A second such effort, META II, was begun in Argentina in 1990, to search the southern sky. META II is still in operation, after an equipment upgrade in 1996.

The follow-on to META was named "BETA", for "Billion-channel Extraterrestrial Assay", and it commenced observation on October 30, 1995. The heart of BETA's processing capability consisted of 63 dedicated fast Fourier transform (FFT) engines, each capable of performing a 2^{22}-point complex FFTs in two seconds, and 21 general-purpose personal computers equipped with custom digital signal processing boards. This allowed BETA to receive 250 million simultaneous channels with a resolution of 0.5 hertz per channel. It scanned through the microwave spectrum from 1.400 to 1.720 gigahertz in eight hops, with two seconds of observation per hop. An important capability of the BETA search was rapid and automatic re-observation of candidate signals, achieved by observing the sky with two adjacent beams, one slightly to the east and the other slightly to the west. A successful candidate signal would first transit the east beam, and then the west beam and do so with a speed consistent with Earth's sidereal rotation rate. A third receiver observed the horizon to veto signals of obvious terrestrial origin. On March 23, 1999, the 26-meter radio telescope on which Sentinel, META and BETA were based was blown over by strong winds and seriously damaged. This forced the BETA project to cease operation.

MOP and Project Phoenix

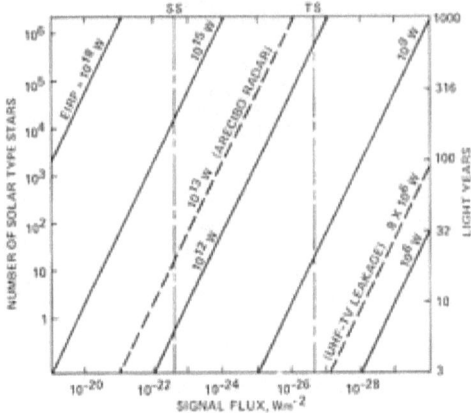

Sensitivity vs range for SETI radio searches. The diagonal lines show transmitters of different effective powers. The x-axis is the sensitivity of the search. The y-axis on the right is the range in light-years, and on the left is the number of Sun-like stars within this range. The vertical line labeled SS is the typical sensitivity achieved by a full sky search, such as BETA above. The vertical line labeled TS is the typical sensitivity achieved by a targeted search such as Phoenix.

In 1978, the NASA SETI program had been heavily criticized by Senator William Proxmire, and funding for SETI research was removed from the NASA budget by Congress in 1981; however, funding was restored in 1982, after Carl Sagan talked with Proxmire and convinced him of the program's value. In 1992, the U.S. government funded an operational SETI program, in the form of the NASA Microwave Observing Program (MOP). MOP was planned as a long-term effort to conduct a general survey of the sky and carry out targeted searches of 800 specific nearby stars. MOP was to be performed by radio antennas associated with the NASA Deep Space Network, as well as the 140-foot (43 m) radio telescope of the National Radio Astronomy Observatory at Green Bank, West Virginia and the 1,000-foot (300 m) radio telescope at the Arecibo Observatory in Puerto Rico. The signals were to be analyzed by spectrum analyzers, each with a capacity of 15 million channels. These spectrum analyzers could be grouped together to obtain greater capacity. Those used in the targeted search had a bandwidth of 1 hertz per channel, while those used in the sky survey had a bandwidth of 30 hertz per channel.

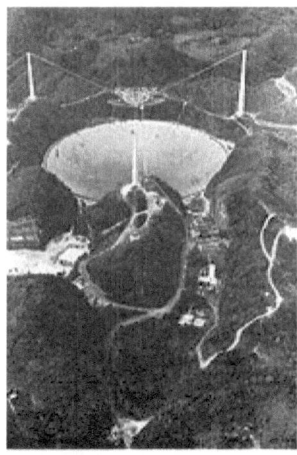

Arecibo Observatory in Puerto Rico with its 300 m (980 ft) dish, one of the world's largest filled-aperture (i.e. full dish) radio telescope, conducts some SETI searches.

MOP drew the attention of the United States Congress, where the program was ridiculed and canceled one year after its start. SETI advocates continued without government funding, and in 1995 the nonprofit SETI Institute of Mountain View, California resurrected the MOP program under the name of Project "Phoenix", backed by private sources of funding. Project Phoenix, under the direction of Jill Tarter, is a continuation of the targeted search program from MOP and studies roughly 1,000 nearby Sun-like stars. From 1995 through March 2004, Phoenix conducted observations at the 64-meter (210 ft) Parkes radio telescope in Australia, the 140-foot (43 m) radio telescope of the National Radio Astronomy Observatory in Green Bank, West Virginia, and the 1,000-foot (300 m) radio telescope at the Arecibo Observatory in Puerto Rico. The project observed the equivalent of 800 stars over the available channels in the frequency range from 1200 to 3000 MHz The search was sensitive enough to pick up transmitters with 1 GW

EIRP to a distance of about 200 light-years. According to Prof. Tarter, in 2012 it costs around "$2 million per year to keep SETI research going at the SETI Institute" and approximately 10 times that to support "all kinds of SETI activity around the world".

Ongoing Radio Searches

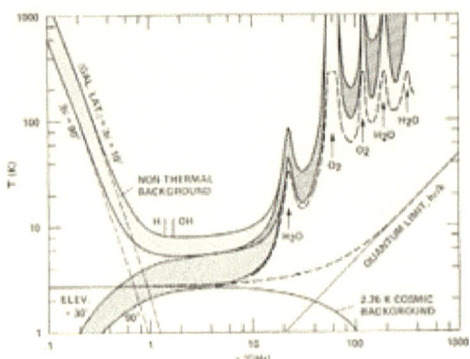

Microwave window as seen by a ground-based system. From NASA report SP-419: SETI – the Search for Extraterrestrial Intelligence

Many radio frequencies penetrate Earth's atmosphere quite well, and this led to radio telescopes that investigate the cosmos using large radio antennas. Furthermore, human endeavors emit considerable electromagnetic radiation as a byproduct of communications such as television and radio. These signals would be easy to recognize as artificial due to their repetitive nature and narrow bandwidths. If this is typical, one way of discovering an extraterrestrial civilization might be to detect artificial radio emissions from a location outside the Solar System.

Many international radio telescopes are currently being used for radio SETI searches, including the Low Frequency Array (LOFAR) in Europe, the Murchison Widefield Array (MWA) in Australia, and the Lovell Telescope in the United Kingdom.

Allen Telescope Array

The SETI Institute collaborated with the Radio Astronomy Laboratory at the Berkeley SETI Research Center to develop a specialized radio telescope array for SETI studies, something like a mini-cyclops array. Formerly known as the One Hectare Telescope (1HT), the concept was renamed the "Allen Telescope Array" (ATA) after the project's benefactor Paul Allen. Its sensitivity would be equivalent to a single large dish more than 100 meters in diameter if completed. Presently, the array under construction has 42 dishes at the Hat Creek Radio Observatory in rural northern California.

The full array (ATA-350) is planned to consist of 350 or more offset-Gregorian radio dishes, each 6.1 meters (20 feet) in diameter. These dishes are the largest producible with commercially available satellite television dish technology. The ATA was planned for a 2007 completion date, at a very modest cost of US$25 million. The SETI Institute provided money for building the ATA while University of California, Berkeley designed the telescope and provided operational funding. The first portion of the array (ATA-42) became operational in October 2007 with 42 antennas. The DSP system planned for ATA-350 is extremely ambitious. Completion of the full 350 element array will depend on funding and the technical results from ATA-42.

ATA-42 (ATA) is designed to allow multiple observers simultaneous access to the interferometer output at the same time. Typically, the ATA snapshot imager (used for astronomical surveys and SETI) is run in parallel with the beam forming system (used primarily for SETI). ATA also supports observations in multiple synthesized pencil beams at once, through a technique known as "multibeaming". Multibeaming provides an effective filter for identifying false positives in SETI, since a very distant transmitter must appear at only one point on the sky.

SETI Institute's Center for SETI Research (CSR) uses ATA in the search for extraterrestrial intelligence, observing 12 hours a day, 7 days a week. From 2007-2015, ATA has identified hundreds of millions of technological signals. So far, all these signals have been assigned the status of noise or radio frequency interference because a) they appear to be generated by satellites or Earth-based transmitters, or b) they disappeared before the threshold time limit of ~1 hour. Researchers in CSR are presently working on ways to reduce the threshold time limit, and to expand ATA's capabilities for detection of signals that may have embedded messages.

Berkeley astronomers used the ATA to pursue several science topics, some of which might have turned up transient SETI signals, until 2011, when the collaboration between the University of California and the SETI Institute was terminated.

CNET published an article and pictures about the Allen Telescope Array (ATA) on December 12, 2008.

In April 2011, the ATA was forced to enter an 8-month "hibernation" due to funding shortfalls. Regular operation of the ATA was resumed on December 5, 2011.

In 2012, new life was breathed into the ATA thanks to a $3.6M philanthropic donation by Franklin Antonio, Co-Founder and Chief Scientist of QUALCOMM Incorporated. This gift supports upgrades of all the receivers on the ATA dishes to have dramatically (2x - 10x from 1–8 GHz) greater sensitivity than before and supporting sensitive observations over a wider frequency range from 1–18 GHz, though initially the radio frequency electronics go to only 12 GHz. As of July 2013, the first of these receivers was installed and proven. Full installation on all 42 antennas is expected in June 2014. ATA is especially well suited to the search for extraterrestrial intelligence SETI and to discovery of astronomical radio sources, such as heretofore unexplained non-repeating, possibly extragalactic, pulses known as fast radio bursts or FRBs.

SERENDIP

SERENDIP (Search for Extraterrestrial Radio Emissions from Nearby Developed Intelligent Populations) is a SETI program launched in 1979 by the Berkeley SETI Research Center. SERENDIP takes advantage of ongoing "mainstream" radio telescope observations as a "piggy-back" or "commensal" program, using large radio telescopes including the NRAO 90m telescope at Green Bank and the Arecibo 305m telescope. Rather than having its own observation program, SERENDIP analyzes deep space radio telescope data that it obtains while other astronomers are using the telescopes.

The most recently deployed SERENDIP spectrometer, SERENDIP V.v, was installed at the Arecibo Observatory in June 2009 and is currently operational. The digital back-end instrument is an FPGA-based 128 million-channel digital spectrometer covering 200 MHz of bandwidth. It takes data commensally with the seven-beam Arecibo L-band Feed Array[43] (ALFA). The program has found around 400 suspicious signals, but there is not enough data to prove that they belong to extraterrestrial intelligence.[44]

Breakthrough Listen

Breakthrough Listen is a ten-year initiative with $100 million funding begun in July 2015 to actively search for intelligent extraterrestrial communications in the universe, in a substantially expanded way, using resources that had not previously been extensively used for the purpose. It has been described as the most comprehensive search for alien communications to date. The science program for Breakthrough Listen is based at Berkeley SETI Research Center, located in the Astronomy Department at the University of California, Berkeley.

Announced in July 2015, the project is observing for thousands of hours every year on two major radio telescopes, the Green Bank Observatory in West Virginia and the Parkes Observatory in Australia. Previously, only about 24 to 36 hours of telescope per year was used in the search for alien life. Furthermore, the Automated Planet Finder at Lick Observatory is searching for optical signals coming from laser transmissions. The massive data rates from the radio telescopes (24 GB/s at Green Bank) necessitated the construction of dedicated hardware at the telescopes to perform the bulk of the analysis. Some of the data are also analyzed by volunteers in the SETI@home distributed computing network. Founder of modern SETI Frank Drake is one of the scientists on the project's advisory committee.

FAST

China's 500-meter Aperture Spherical Telescope (FAST) lists *detecting interstellar communication signals* as part of its science mission. It is funded by the National Development and Reform Commission (NDRC) and managed by the National Astronomical observatories (NAOC) of Chinese Academy of Sciences (CAS). FAST consists of a fixed 500 m (1,600 ft) diameter spherical dish constructed in a natural depression sinkhole caused by karst processes in the region. It is the world's largest filled-aperture radio telescope. According to its website, FAST could search out to 28 light-years, and would be able to reach 1400 stars. If the transmitter's radiated power is increased to 1000,000 MW, FAST would be able to reach one million stars. This is compared to the Arecibo 305-meter telescope detection distance of 18 light-years.

Community SETI projects

SETI@home was conceived by David Gedye along with Craig Kasnoff and is a popular volunteer distributed computing project that was launched by the Berkeley SETI Research Center at the University of California, Berkeley, in May 1999. It was originally funded by The Planetary Society and Paramount Pictures, and later by the state of California. The project is run by director David P. Anderson and chief scientist Dan Werthimer. Any individual can become involved with SETI research by downloading the Berkeley

Open Infrastructure for Network Computing (BOINC) software program, attaching to the SETI@home project, and allowing the program to run as a background process that uses idle computer power. The SETI@home program itself runs signal analysis on a "work unit" of data recorded from the central 2.5 MHz wide band of the SERENDIP IV instrument. After computation on the work unit is complete, the results are then automatically reported back to SETI@home servers at University of California, Berkeley. By June 28, 2009, the SETI@home project had over 180,000 active participants volunteering a total of over 290,000 computers. These computers give SETI@home an average computational power of 617 teraflops. In 2004 radio source SHGb02+14a set off speculation in the media that a signal had been detected but researchers noted the frequency drifted rapidly and the detection on three SETI@home computers fell within random chance.

As of 2010, after 10 years of data collection, SETI@home has listened to that one frequency at every point of over 67 percent of the sky observable from Arecibo with at least three scans (out of the goal of nine scans), which covers about 20 percent of the full celestial sphere.

SETI Net

SETI Net is a private search system created by a single individual. It is closely affiliated with the SETI League and is one of the project Argus stations (DM12jw).

The SETI Net station consists of off-the-shelf, consumer-grade electronics to minimize cost and to allow this design to be replicated as simply as possible. It has a 3-meter parabolic antenna that can be directed in azimuth and elevation, an LNA that covers the 1420 MHz spectrum, a receiver to reproduce the wideband audio, and a standard personal computer as the control device and for deploying the detection algorithms.

The antenna can be pointed and locked to one sky location, enabling the system to integrate on it for long periods. Currently the Wow! signal area is being monitored when it is above the horizon. All search data are collected and made available on the Internet archive.

SETI Net started operation in the early 1980s to learn about the science of the search and has developed several software packages for the amateur SETI community. It has provided an astronomical clock, a file manager to keep track of SETI data files, a spectrum analyzer optimized for amateur SETI, remote control of the station from the Internet, and other packages.

The SETI League and Project Argus

Founded in 1994 in response to the United States Congress cancellation of the NASA SETI program, The SETI League, Inc. is a membership-supported nonprofit organization with 1,500 members in 62 countries. This grass-roots alliance of amateur and professional radio astronomers is headed by executive director emeritus H. Paul Shuch, the engineer credited with developing the world's first commercial home satellite TV receiver. Many SETI League members are licensed radio amateurs and microwave experimenters. Others are digital signal processing experts and computer enthusiasts.

The SETI League pioneered the conversion of backyard satellite TV dishes 3 to 5 m (10–16 ft) in diameter into research-grade radio telescopes of modest sensitivity. The organization concentrates on coordinating a global network of small, amateur-built radio telescopes under Project Argus, an all-sky survey seeking to achieve real-time coverage of the entire sky. Project Argus was conceived as a continuation of the all-sky survey component of the late NASA SETI program (the targeted search having been continued by the SETI Institute's Project Phoenix). There are currently 143 Project Argus radio telescopes operating in 27 countries. Project Argus instruments typically exhibit sensitivity on the order of 10^{-23} Watts/square meter, or roughly equivalent to that achieved by the Ohio State University Big Ear radio telescope in 1977, when it detected the landmark "Wow!" candidate signal.

The name "Argus" derives from the mythical Greek guard-beast who had 100 eyes, and could see in all directions at once. In the SETI context, the name has been used for radio telescopes in fiction (Arthur C. Clarke, *"Imperial Earth"*; Carl Sagan, *"Contact"*), was the name initially used for the NASA study ultimately known as "Cyclops," and is the name given to an omnidirectional radio telescope design being developed at the Ohio State University.

Optical Experiments

While most SETI sky searches have studied the radio spectrum, some SETI researchers have considered the possibility that alien civilizations might be using powerful lasers for interstellar communications at optical wavelengths. The idea was first suggested by R. N. Schwartz and Charles Hard Townes in a 1961 paper published in the journal *Nature* titled "Interstellar and Interplanetary Communication by Optical Masers". However, the 1971 Cyclops study discounted the possibility of optical SETI, reasoning that construction of a laser system that could outshine the bright central star of a remote star system would be too difficult. In 1983, Townes published a detailed study of the idea in the United States journal *Proceedings of the National Academy of Sciences*, which was met with widespread agreement by the SETI community.

There are two problems with optical SETI. The first problem is that lasers are highly "monochromatic", that is, they emit light only on one frequency, making it troublesome to figure out what frequency to look for. However, emitting light in narrow pulses results in a broad spectrum of emission; the spread in frequency becomes higher as the pulse width becomes narrower, making it easier to detect an emission.

The other problem is that while radio transmissions can be broadcast in all directions, lasers are highly directional. Interstellar gas and dust are almost transparent to near infrared, so these signals can be seen from greater distances, but the extraterrestrial laser signals would need to be transmitted in the direction of Earth to be detected.

Optical SETI supporters have conducted paper studies of the effectiveness of using contemporary high-energy lasers and a ten-meter diameter mirror as an interstellar beacon. The analysis shows that an infrared pulse from a laser, focused into a narrow beam by such a mirror, would appear thousands of times brighter than the Sun to a distant civilization in the beam's line of fire. The Cyclops study proved incorrect in suggesting a laser beam would be inherently hard to see.

Such a system could be made to automatically steer itself through a target list, sending a pulse to each target at a constant rate. This would allow targeting of all Sun-like stars within 100 light-years. The studies have also described an automatic laser pulse detector system with a low-cost, two-meter mirror made of carbon composite materials, focusing on an array of light detectors. This automatic detector system could perform sky surveys to detect laser flashes from civilizations attempting contact.

Several optical SETI experiments are now in progress. A Harvard-Smithsonian group that includes Paul Horowitz designed a laser detector and mounted it on Harvard's 155 centimeters (61 inches) optical telescope. This telescope is currently being used for a more conventional star survey, and the optical SETI survey is "piggybacking" on that effort. Between October 1998 and November 1999, the survey inspected about 2,500 stars. Nothing that resembled an intentional laser signal was detected, but efforts continue. The Harvard-Smithsonian group is now working with Princeton University to mount a similar detector system on Princeton's 91-centimeter (36-inch) telescope. The Harvard and Princeton telescopes will be "ganged" to track the same targets at the same time, with the intent being to detect the same signal in both locations as a means of reducing errors from detector noise.

The Harvard-Smithsonian group is now building a dedicated all-sky optical survey system along the lines of that described above, featuring a 1.8-meter (72-inch) telescope. The new optical SETI survey telescope is being set up at the Oak Ridge Observatory in Harvard, Massachusetts.

The University of California, Berkeley, home of SERENDIP and SETI@home, is also conducting optical SETI searches through the NIROSETI program. It is being directed by Geoffrey Marcy, an extrasolar planet hunter, and it involves examination of records of spectra taken during extrasolar planet hunts for a continuous, rather than pulsed, laser signal. This survey uses the Nickel 1-m telescope at the Lick Observatory, situated on the summit of Mount Hamilton, east of San Jose, California, USA. The other Berkeley optical SETI effort is being pursued by the Harvard-Smithsonian group and is being directed by Dan Werthimer of Berkeley, who built the laser detector for the Harvard-Smithsonian group. This survey uses a 76-centimeter (30-inch) automated telescope at Leuschner Observatory and an older laser detector built by Werthimer.

In May 2017, astronomers reported studies related to laser light emissions from stars, as a way of detecting technology-related signals from an alien civilization. The reported studies included KIC 8462852, an oddly dimming star in which its unusual starlight fluctuations may be the result of interference by an artificial megastructure, such as a Dyson swarm, made by such a civilization. No evidence was found for technology-related signals from KIC 8462852 in the studies.

Gamma-ray Bursts

Gamma-ray bursts (GRBs) are candidates for extraterrestrial communication. These high-energy bursts are observed about once per day and originate throughout the observable universe. SETI currently omits gamma ray frequencies in their monitoring and analysis because they are absorbed by the Earth's atmosphere and difficult to detect with ground-based receivers. In addition, the wide burst bandwidths pose a serious analysis challenge for modern digital signal processing systems. Still, the continued mysteries surrounding gamma-ray bursts have encouraged hypotheses invoking extraterrestrials. John A. Ball from the MIT Haystack Observatory suggests that an advanced civilization that has reached a technological singularity would be capable of transmitting a two-millisecond pulse encoding 1×10^{18} bits of information. This is "comparable to the estimated total information content of Earth's biosystem—genes and memes and including all libraries and computer media".

Search for Extraterrestrial Artifacts

The possibility of using interstellar messenger probes in the search for extraterrestrial intelligence was first suggested by Ronald N. Bracewell in 1960 (see Bracewell probe), and the technical feasibility of this approach was demonstrated by the British Interplanetary Society's starship study Project Daedalus in 1978. Starting in 1979, Robert Freitas advanced arguments for the proposition that physical space-probes are a superior mode of interstellar communication to radio signals. See Voyager Golden Record.

In recognition that any sufficiently advanced interstellar probe in the vicinity of Earth could easily monitor the terrestrial Internet, Invitation to ETI was established by Prof. Allen Tough in 1996, as a Web-based SETI experiment inviting such spacefaring probes to establish contact with humanity. The project's 100 Signatories includes prominent physical, biological, and social scientists, as well as artists, educators, entertainers, philosophers and futurists. Prof. H. Paul Shuch, executive director emeritus of The SETI League, serves as the project's Principal Investigator.

Inscribing a message in matter and transporting it to an interstellar destination can be enormously more energy efficient than communication using electromagnetic waves if delays larger than light transit time can be tolerated. That said, for simple messages such as "hello," radio SETI could be far more efficient. If energy requirement is used as a proxy for technical difficulty, then a solar centric Search for Extraterrestrial Artifacts (SETA) may be a useful supplement to traditional radio or optical searches.

Much like the "preferred frequency" concept in SETI radio beacon theory, the Earth-Moon or Sun-Earth libration orbits might therefore constitute the most universally convenient parking places for automated extraterrestrial spacecraft exploring arbitrary stellar systems. A viable long-term SETI program may be founded upon a search for these objects.

In 1979, Freitas and Valdes conducted a photographic search of the vicinity of the Earth-Moon triangular libration points L_4 and L_5, and of the solar-synchronized positions in the associated halo orbits, seeking possible orbiting extraterrestrial interstellar probes, but found nothing to a detection limit of about 14th magnitude. The authors conducted a second, more comprehensive photographic search for probes in 1982 that examined the five Earth-Moon Lagrangian positions and included the solar-synchronized positions in the stable L4/L5 libration orbits, the potentially stable nonplanar orbits near L1/L2, Earth-Moon L_3, and also L_2 in the Sun-Earth system. Again no extraterrestrial probes were found to limiting magnitudes of 17–19th magnitude near L3/L4/L5, 10–18th magnitude for L_1/L_2, and 14–16th magnitude for Sun-Earth L_2.

In June 1983, Valdes and Freitas[80] used the 26 m radiotelescope at Hat Creek Radio Observatory to search for the tritium hyperfine line at 1516 MHz from 108 assorted astronomical objects, with emphasis on 53 nearby stars including all visible stars within a 20 light-year radius. The tritium frequency was deemed highly attractive for SETI work because (1) the isotope is cosmically rare, (2) the tritium hyperfine line is centered in the SETI waterhole region of the terrestrial microwave window, and (3) in addition to beacon signals, tritium hyperfine emission may occur as a byproduct of extensive nuclear fusion energy production by extraterrestrial civilizations. The wideband- and narrowband-channel observations achieved sensitivities of $5–14 \times 10^{-21}$

W/m²/channel and 0.7-2 x 10^{-24} W/m²/channel, respectively, but no detections were made.

Technosignatures

Technosignatures, including all signs of technology except for the interstellar radio messages that define traditional SETI, are a recent avenue in the search for extraterrestrial intelligence. Technosignatures may originate from various sources, from megastructures such as Dyson spheres and space mirrors or space shaders to the atmospheric contamination created by an industrial civilization, or city lights on extrasolar planets, and may be detectable in the future with large hyper telescopes.[83]

Technosignatures can be divided into three broad categories: astroengineering projects, signals of planetary origin, and spacecraft within and outside the Solar System.

An astroengineering installation such as a Dyson sphere, designed to convert all of the incident radiation of its host star into energy, could be detected through the observation of an infrared excess from a solar analog star, or by the star's apparent disappearance in the visible spectrum over several years. After examining some 100,000 nearby large galaxies, a team of researchers has concluded that none of them display any obvious signs of highly advanced technological civilizations.

Another hypothetical form of astroengineering, the Shkadov thruster, moves its host star by reflecting some of the star's light back on itself, and would be detected by observing if its transits across the star abruptly end with the thruster in front. Asteroid mining within the Solar System is also a detectable technosignature of the first kind.

Individual extrasolar planets can be analyzed for signs of technology. Avi Loeb of the Harvard-Smithsonian Center for Astrophysics has proposed that persistent light signals on the night side of an exoplanet can be an indication of the presence of cities and an advanced civilization. In addition, the excess infrared radiation and chemicals produced by various industrial processes or terraforming efforts may point to intelligence.

Clearly, light and heat detected from planets need to be distinguished from natural sources to conclusively prove the existence of civilization on a planet. However, as argued by the Colossus team, a civilization heat signature should be within a "comfortable" temperature range, like terrestrial urban heat islands, i.e. only a few degrees warmer than the planet itself. In contrast, such natural sources as wild fires, volcanoes, etc. are significantly hotter, so they will be well distinguished by their maximum flux at a different wavelength.

Extraterrestrial craft are another target in the search for technosignatures. Magnetic sail interstellar spacecraft should be detectable over thousands of light-years of distance through the synchrotron radiation they would produce through interaction with the interstellar medium; other interstellar spacecraft designs may be detectable at more modest distances. In addition, robotic probes within the Solar System are also being sought out with optical and radio searches.

For a sufficiently advanced civilization, hyper energetic neutrinos from Planck scale accelerators should be detectable at many Mpc.

Post Detection Disclosure Protocol

The International Academy of Astronautics (IAA) has a long-standing SETI Permanent Study Group (SPSG, formerly called the IAA SETI Committee), which addresses matters of SETI science, technology, and international policy. The SPSG meets in conjunction with the International Astronautical Congress (IAC) held annually at different locations around the world, and sponsors two SETI Symposia at each IAC. In 2005, the IAA established the SETI: Post-Detection Science and Technology Task group (Chairman, Professor Paul Davies) "to act as a Standing Committee to be available to be called on at any time to advise and consult on questions stemming from the discovery of a putative signal of extraterrestrial intelligent (ETI) origin."

However, the protocols mentioned apply only to radio SETI rather than for METI (Active SETI). The intention for METI is covered under the SETI charter "Declaration of Principles Concerning Sending Communications with Extraterrestrial Intelligence".

On October 2000 astronomers Iván Almár and Jill Tarter presented a paper to The SETI Permanent Study Group in Rio de Janeiro, Brazil which proposed a scale (modelled after the Torino scale) which is an ordinal scale between zero and ten that quantifies the impact of any public announcement regarding evidence of extraterrestrial intelligence; the **Rio scale** has since inspired the 2005 San Marino Scale (in regard to the risks of transmissions from Earth) and the 2010 London Scale (in regard to the detection of extraterrestrial life)

The SETI Institute does not officially recognize the Wow! signal as of extraterrestrial origin (as it was unable to be verified). The SETI Institute has also publicly denied that the candidate signal Radio source SHGb02+14a is of extraterrestrial origin though full details of the signal, such as its exact location have never been disclosed to the public. Although other volunteering projects such as Zooniverse credit users for discoveries, there is currently no crediting or early notification by SETI@Home following the discovery of a signal.

Some people, including Steven M. Greer, have expressed cynicism that the general public might not be informed in the event of a genuine discovery of extraterrestrial intelligence due to significant vested interests. Some, such as Bruce Jakosky have also argued that the official disclosure of extraterrestrial life may have far reaching and as yet undetermined implications for society, particularly for the world's religions.

Active SETI

Active SETI, also known as messaging to extraterrestrial intelligence (METI), consists of sending signals into space in the hope that they will be picked up by an alien intelligence.

Realized interstellar radio message projects

In November 1974, a largely symbolic attempt was made at the Arecibo Observatory to send a message to other worlds. Known as the Arecibo Message, it was sent towards the globular cluster M13, which is 25,000 light-years from Earth. Further IRMs Cosmic Call, Teen Age Message, Cosmic Call 2, and A Message From Earth were transmitted in 1999, 2001, 2003 and 2008 from the Evpatoria Planetary Radar.

Debate

Physicist Stephen Hawking, in his book *A Brief History of Time*, suggests that "alerting" extraterrestrial intelligences to our existence is foolhardy, citing mankind's history of treating his fellow man harshly in meetings of civilizations with a significant technology gap. He suggests, in view of this history, that we "lay low". In one response to Hawking, in September 2016, astronomer Seth Shostak, allays such concerns. Astronomer Jill Tarter also disagrees with Hawking, arguing that aliens developed and long-lived enough to communicate and travel across interstellar distances would have evolved a cooperative and less violent intelligence. She does think it is too soon for humans to attempt active SETI and that humans should be more advanced technologically first but keep listening in the meantime.

The concern over METI was raised by the science journal *Nature* in an editorial in October 2006, which commented on a recent meeting of the International Academy of Astronautics SETI study group. The editor said, "It is not obvious that all extraterrestrial civilizations will be benign, or that contact with even a benign one would not have serious repercussions" (Nature Vol 443 12 October 06 p 606). Astronomer and science fiction author David Brin have expressed similar concerns.

Richard Carrigan, a particle physicist at the Fermi National Accelerator Laboratory near Chicago, Illinois, suggested that passive SETI could also be dangerous and that a signal released onto the Internet could act as a computer virus. Computer security expert Bruce Schneier dismissed this possibility as a "bizarre movie-plot threat".

To lend a quantitative basis to discussions of the risks of transmitting deliberate messages from Earth, the SETI Permanent Study Group of the International Academy of Astronautics adopted in 2007 a new analytical tool, the San Marino Scale. Developed by Prof. Ivan Almar and Prof. H. Paul Shuch, the scale evaluates the significance of transmissions from Earth as a function of signal intensity and information content. Its adoption suggests that not all such transmissions are equal, and each must be evaluated separately before establishing blanket international policy regarding active SETI.

However, some scientists consider these fears about the dangers of METI as panic and irrational superstition; see, for example, Alexander L. Zaitsev's papers. Biologist João Pedro de Magalhães also proposed in 2015 transmitting an invitation message to any extraterrestrial intelligences watching us already in the context of the Zoo Hypothesis and inviting them to respond, arguing this would not put us in any more danger than we are already if the Zoo Hypothesis is correct.

On 13 February 2015, scientists (including Geoffrey Marcy, Seth Shostak, Frank Drake, Elon Musk and David Brin) at a convention of the American Association for the Advancement of Science, discussed Active SETI and whether transmitting a message to possible intelligent extraterrestrials in the Cosmos was a good idea; one result was a statement, signed by many, that a "worldwide scientific, political and humanitarian discussion must occur before any message is sent". On 28 March 2015, a related essay was written by Seth Shostak and published in *The New York Times*.

Breakthrough Message

The Breakthrough Message program is an open competition announced in July 2015 to design a digital message that could be transmitted from Earth to an extraterrestrial civilization, with a US$1,000,000 prize pool. The message should be "representative of humanity and planet Earth". The program pledges "not to transmit any message until there has been a wide-ranging debate at high levels of science and politics on the risks and rewards of contacting advanced civilizations".

Chapter 5

How Do Scientists Search for Extraterrestrials?

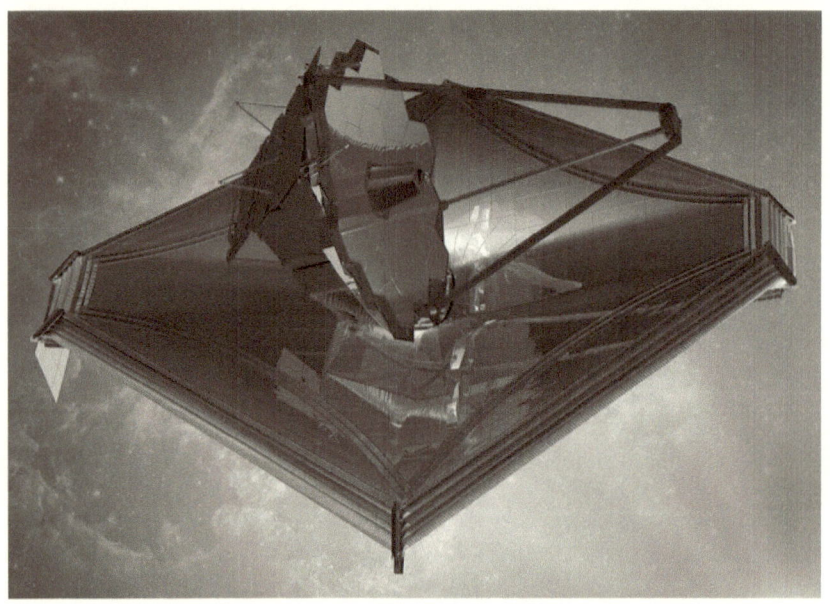

The James Webb Space Telescope, expected to launch in 2018, will offer views of distant galaxies in unprecedented detail, and could reveal undiscovered Earth-like planets.

Credit: Northrop Grumman

Human civilizations dating back thousands of years left behind structures and records documenting their studies of the stars as they sought to chart the seasons, help travelers find their way and interpret the world around them. Stargazers among the ancient Greeks, Maya, Egyptians, Middle Easterners and Asians likely also pondered if there were other planets like ours among those distant points of light — and if so, what might live there.

Over the last century, science-fiction storytellers have used books, movies, comics and television to speculate at great length about contact with creatures from other worlds — to our benefit and our detriment. These creatures have been imagined as sometimes benevolent and sometimes bloodthirsty, and they have come in a wide range of shapes and sizes — from inquisitive "little green men" to human-parasitizing, chest-bursting Xenomorphs in the "Alien" movie franchise.

Present-day astronomers have likewise been probing this question, using sophisticated equipment to listen farther and peer deeper into the universe than ever before, to find evidence of our cosmic neighbors. From detecting unexplained radio signals to investigating the atmospheres and liquid water on distant worlds — how are scientists searching for signs of extraterrestrial life? [Greetings, Earthlings! 8 Ways Aliens Could Contact Us]

For an alien-seeking scientist, "life" means any living form — including microbes, astronomer Mercedes López-Morales, at the Harvard-Smithsonian Center for Astrophysics in Cambridge, Massachusetts, told Live Science.

But even the smallest microbe living on a distant exoplanet — a planet orbiting a star other than our sun — could still broadcast a chemical signal that would be visible to sensitive telescopes, in the form of atmospheric gases that probably wouldn't be there in the absence of life, López-Morales explained.

"Life affects the atmosphere of a planet," she said. "You have gases that are only there because they are constantly being replenished by something — otherwise, they would react with other gases and disappear. For that gas or that molecule to be in the atmosphere of a planet, it must have some mechanism that is continuously producing it," López-Morales said.

One of the atmospheric gases astronomers are searching for in exoplanets is oxygen, which is plentiful in Earth's atmosphere because it is continuously being replaced by plants through photosynthesis.

However, the presence of unusual atmospheric gases doesn't necessarily mean that something living is generating them, López-Morales added.

"Sulfur molecules, for example, could come from active volcanoes," she explained. "For oxygen, there are at least two or three ways to produce it that involve irradiation in the ultraviolet light coming from stars. But we know that oxygen appeared on Earth because life appeared on Earth," she said.

An artist's concept imagines the surface of the exoplanet TRAPPIST-1f, located in the TRAPPIST-1 system in the constellation Aquarius, where liquid water could harbor extraterrestrial life.

Credit: NASA/JPL-Caltech/T-Pyle (IPAC)

Of course, even if these chemical signatures can be detected, there's no way to tell what forms of life are producing the signal, Sara Seager, an astrophysicist and planetary scientist at MIT, told Live Science in an email.

And what type of exoplanet is a good candidate for life? Our familiarity with our own world nudges efforts toward those that resemble Earth — "a rocky planet with a thin atmosphere with surface water," Seager said.

"Right now, we can tell — for some planets — if they are rocky, based on planet size and planet mass, which gives average density. But we can't yet tell if a planet has liquid water," she said. [A Field Guide to Alien Planets]

Location, location, location

What else makes an exoplanet a promising candidate? "Anything close by," López-Morales told Live Science. For an astronomer, that means less than 30 light-years away, which would enable humans to visit the world where life was detected, she said. (One light-year is about 5.9 trillion miles, or 9.5 trillion kilometers.)

"Eventually, I hope humans will have the technology to get that far, within a reasonable number of years. So, for us, the holy grail is to find something within 30 light-years of Earth," she said.

Scientists are also investigating worlds within our own solar system — such as the Saturn moons Titan and Enceladus — which are close enough to be visited by probes that can collect samples and capture images. Several NASA missions are also looking closely at Mars, which once had abundant liquid water on its surface, and where brackish water still flows today, researchers announced in 2015.

This May 11, 2016, self-portrait of NASA's Curiosity Mars rover shows the vehicle at the "Okoruso" drilling site on lower Mount Sharp's "Naukluft Plateau." Scientists want to send Curiosity higher on Mount Sharp, to look for evidence of liquid water.

Credit: NASA/JPL-Caltech/MSSS

"Humans are creatures that want to know — where we came from, where we're going, how we appeared on Earth," López-Morales said. "Our research might start providing answers to that." [FAQ: Significance of Liquid Water on Mars]

Radio signals

But scientists aren't just looking for signs of extraterrestrial life — they're also listening for them.

For more than two decades, SETI, the Search for Extraterrestrial Intelligence Institute, has conducted research to understand the origins of life in the universe, and to detect and analyze evidence of life emanating from places other than Earth. This effort includes investigations of microbial life within our solar system, such as on the surface of Mars or under the icy crust of Jupiter's moon Europa. SETI scientists are also monitoring the universe for signals in light or radio wavelengths that originate far away and could be signs of technologically advanced alien life, SETI explains on its website.

At SETI, astronomers use the Allen Telescope Array (ATA) of 42 radio antennas to "listen" for signals over a range of radio frequencies, tuned to "hear" the regions around 20,000 red dwarf stars (a broad term describing stars smaller than our sun and in a certain spectral range) that are closest to Earth, Seth Shostak, a senior astronomer at the SETI Institute, told Live Science.

The radio astronomy facility at Owens Valley Observatory in Owens Valley California, which has been used by the SETI program.

Credit: Harun Mehmedinovic and Gavin Heffernan

Investigating red dwarf stars for life-supporting worlds is a relatively recent development at SETI. In the past, stars that were more like our own sun — a yellow dwarf — were thought to be the most likely candidates to host planets harboring life. But over the last few decades, astronomers have determined that many red dwarf stars host planets that could be at the right distance from the star to be habitable, according to Shostak.

"That's something we didn't know when we started," he said.

And SETI radio-signal monitoring is accelerating, as telescopes become more sensitive and technological developments increase the number of radio channels and locations in the sky that can be studied at once, Shostak explained.

"Until now, the total number of star systems that have been looked at carefully over a wide range of the radio dial is measured in the thousands. In the next 20 years, with new technology, you could increase that number to maybe a million," he said. [4 Places Where Alien Life May Lurk in the Solar System]

An alien megastructure?

Shostak also reviews images of alleged alien spacecraft sent to him by hopeful photographers, he told Live Science. A photographer himself, Shostak said that he invariably identifies all the purported "UFO" sightings as tricks of the light or internal reflections in the camera lens — much to the dismay of the observers.

"That never makes them happy," he said.

But even among astronomers, unusual observations can sometimes turn the conversation toward the likelihood of alien technology.

In 2015, when scientists discovered the star KIC 8462852 — also known as Tabby's Star, located more than 1,400 light-years from Earth — they were puzzled by repeated and significant dips in its brightness that took place over several years. During the dips, the star dimmed by as much as 22 percent, far more than could be caused by an orbiting planet passing in front of the star, Shostak said.

In short, the star was "really weird," Tabetha Boyajian, lead author of a study about the star and a researcher at Yale University, told the Atlantic in October of that year.

One possible explanation suggested by some experts was an "alien megastructure," an enormous array orbiting KIC 8462852, built by a hypothetical alien civilization advanced enough to possess technology capable of drawing power from a star. Such a construct could — in theory — periodically block visible light and make the star appear dramatically dimmer when seen from Earth, Space.com reported in 2015.

However, there is no data to actively support this hypothesis. In fact, on all fronts, evidence of any extraterrestrial presence — within our own solar system or beyond its boundaries — remains elusive. But scientists seeking life on other worlds are undaunted by the ongoing challenge, Shostak told Live Science.

"The search should continue, simply because it's a very interesting question," he said.

"Is Earth special? Is it the only place around with intelligent life? That would be remarkable — but it's just as remarkable to find you're not the only kid on the block. That's something that would change our view of ourselves forever," he said.

If intelligent, communicating civilizations exist in the Milky Way, how can we learn that they are there? While there are many reports of UFOs in the popular media, to date there has been no credible evidence that any alien civilization has ever visited the Earth. Since the distances to the nearest stars are a few light years or more, and since our current technology only allows us to build ships that achieve velocities that still require years to reach Pluto, the stars are unreachable to us. Even if other civilizations might be capable of building ships that can fly much faster, a round trip to a star 20 light years away is at the very least 40 years, and likely much longer than that. Since physical travel between Earth and any nearby stars is improbable because of the lengths of time involved, if we are to find other civilizations in the Milky Way, we expect it will be by communication using light (that is, radio waves or optical light) rather than direct visits.

Light travels faster than any other means of communication, so a sufficiently advanced civilization may try to directly communicate with other civilizations using light. Beyond direct, purposeful communication, though, our planet is broadcasting signals out into space every day in the form of our radio and TV broadcasts. That is, when we broadcast radio signals around the world for you to listen to in your car, those same signals also travel through space, and so any civilization with a sophisticated enough detector can receive, say, the "I Love Lucy" show from decades ago. By the same logic, if we try, we should be able to detect signals sent directly to us from a distant civilization, or if they also use transmitters to transmit radio or TV type signals, we could detect those signals, too. However, the signal from a radio transmitter dilutes as it moves farther and farther from Earth, so the radio telescopes a distant civilization must have to detect TV or radio signals from Earth would have to dwarf our most powerful radio telescopes on Earth.

If you return to the lesson on the electromagnetic spectrum and review, there are a few considerations that we or another civilization might want to consider when deciding how to communicate from planet to planet:

- **Cost:** Radio photons carry less energy than say gamma-rays, so it is cheaper to generate radio signals than gamma-ray signals. So, we expect that radio waves are the most efficient way to communicate over large distances.
- **Background:** The Milky Way contains many objects that give off light from radio through gamma rays, and so we want to choose a wavelength of light that will not be swamped by the Milky Way or absorbed as it travels through the interstellar medium.

Since we cannot know ahead of time anything about other civilizations that may be listening for signals from us or who are trying to communicate with us, the best that we can do is take educated guesses at how we might communicate. Scientists who have been pursuing **Search for Extraterrestrial Intelligence** or SETI research have been, since the 1960s, using radio telescopes to search for signals from other civilizations. These searches have concentrated on a region in the radio part of the spectrum known as the **water hole**. In a part of the radio spectrum where the emission from the Galaxy and Earth's atmosphere is at a minimum, there is a wavelength associated with emission from Hydrogen (H) and another with emission from hydroxyl (OH). Since H+OH produces water, this part of the spectrum is referred to as the water hole. The assumption is that since this is a part of the spectrum that many astronomers already study and because the background is very low, it is a logical place for a distant civilization to try to communicate with us. Many of the SETI experiments that have been conducted over the years have tuned their radio telescopes to this part of the spectrum.

The next logical question is, if astronomers have been searching the water hole for a signal from another civilization, has one ever been received? The answer is maybe! In one of the earliest SETI experiments, the "Big Ear" radio telescope detected a signal that is now known as the **"Wow!" Signal**. The Wow Signal has the appearance of a real SETI signal, but it was never able to be independently verified:

```
                    1           2           1    4   3
                    1   16      1       1        1
                    1   11      1       1       11   1
                        1                    3    1
                       6  2                 31
                    1E 24    3   12    1   21    1
        Wow!         Q  1    6  1  2    1    1         1
                    U  1                 3   7        1
                    2 J 1    31  3  11 1    11   1  1
                    5 1                      1   1
                      14     1      1 1 3        2   11
                    1  3     1       1       1
                    1  4             1      1 1     11
                       4     1  1    1   1 1        1 1 1
                       1                 1        2  1
                    1  1     1             11       1
                       1             1              1 4
```

Handwritten note of "wow" near a radio signal detected by the Big Ear observatory

The translation of the numbers on the chart is that each represents the intensity of the signal above the background. The group that reads "6EQUJ5" corresponds to a strong signal that peaks at 30 times the intensity of the background. This is precisely the type of signal that SETI researchers expect to come from an alien civilization broadcasting a radio signal to Earth. Researchers did rule out that this was a terrestrial signal, and no known source of interference was ever discovered that can account for the strength of this signal. To be certain that this is a true SETI contact, though, researchers want verification by observing a repeat of the signal coming from the same part of the sky. Although several searches for a repeat signal were undertaken, none was ever successful.

There have been many different radio SETI searches using the Arecibo radio telescope, the National Radio Astronomy 140-foot radio telescope, the Big Ear telescope, and others. However, researchers have also proposed that optical light may be another option for communication. One thing that SETI researchers consider is how likely it is that an intelligent civilization would be able to generate a powerful signal and would use their resources to do so—after all, would you be willing to put your tax dollars into a device to beam a signal to a planet in case there is life there? If it's cheap or free, you might be persuaded, but if it is expensive, fewer people are likely to see the benefits of such an experiment. So, part of the argument presented by SETI researchers is that, while you can assume that an advanced civilization may be better at generating high power light beams (optical or radio) than we are, they will still want to send signals out to other planets using as few resources (that is, energy) as possible. So, another possibility besides beaming radio signals in the water hole region of the spectrum is that they could shine a pulse of laser light in our direction. These pulses can be *very* bright, but if they are sent in short bursts, they don't use much energy. So "optical SETI" searches are being undertaken to look for short bursts of light from nearby stars.

Want to learn more?

The SETI Institute (link is external) maintains a repository of resources related to the search for extraterrestrial intelligence. They have an excellent history of past SETI projects (link is external). Both the SETI Institute and I recommend watching the movie "Contact" (link is external), based on Carl Sagan's book. Contact gives an accurate depiction of SETI research.

More recently, astronomers, including several at Penn State, have been conducting other searches using different techniques that may reveal intelligent civilizations. One idea concerns "Dyson spheres", that is giant, artificial structures that a civilization might build around a star to capture most of that star's energy to power their civilization. These types of artificial structures should give off waste heat, and therefore, they might be detectable in the infrared. A highly detailed overview of a search for this type of waste heat is posted as a blog by Penn State Professor Jason Wright (link is external). Prof. Wright has also been involved with the study of a very odd star discovered in Kepler Data that is known as "Tabby's Star". In this case, it has been suggested that the highly unusual light curve for this star may be explained not by a planet transiting in front of the star, but an artificial alien "megastructure" that may be like a Dyson sphere transiting in front of the star. It will be very interesting to see what eventual explanation we uncover for this unusual star.

Chapter 6

Using Artificial Intelligence to Search for Extraterrestrial Intelligence

The Machine Learning 4 SETI Code Challenge (ML4SETI), created by the SETI Institute and IBM, was completed on July 31st 2017. Nearly 75 participants, with a wide range of backgrounds from industry and academia, worked in teams on the project. The top team achieved a signal classification accuracy of 95%. The code challenge was sponsored by IBM, Nimbix Cloud, Skymind, Galvanize, and The SETI League.

The ML4SETI project challenged participants to build a machine-learning model to classify different signal types observed in radio-telescope data for the search for extra-terrestrial intelligence (SETI). Seven classes of signals were simulated (and thus, labeled), with which citizen scientists trained their models. We then measured the performance of these models with tests sets to determine a winner of the code challenge. The results were remarkably accurate signal classification models. The models from the top teams, using deep learning techniques, attained nearly 95% accuracy in signals from the test set, which included some signals with very low amplitudes. These models may soon be used in daily SETI radio signal research.

Three of the 42 offset Gregorian, 6-meter dishes that make up the Allen Telescope Array at the Hat Creek Radio Observatory in northern California.

Deep learning models trained for signal classification may significantly impact how SETI research is conducted at the Allen Telescope Array, where the SETI Institute conducts its radio-signal search. More robust classification should allow researchers to improve the efficiency of observing each star system and allow for new ways to implement their search.

Brief explanation of SETI data and its acquisition

To understand the code challenge and exactly how it will help SETI research, an understanding of how the SETI Institute operates is needed. In this section, we'll briefly go over the data acquisition of real SETI data from 2013–2015, the real-time analysis, and how it has been analyzed later in the context of the SETI+IBM collaboration. Some of this information can be found on the SETI Institute's public SETI Quest page.

Time-Series radio signals

The Allen Telescope Array is an array of 42 six-meter-diameter dishes that observe radio signals in the 1–10 GHz range. By combining the signals from different dishes, in a process called "beamforming", observations of radio signals from very small windows of the sky about specific stellar systems are made. At the ATA, three separate beams may be observed simultaneously and are used together to make decisions about the likelihood of observing intelligent signals. On the SETIQuest page, one can see the current observations in real-time.

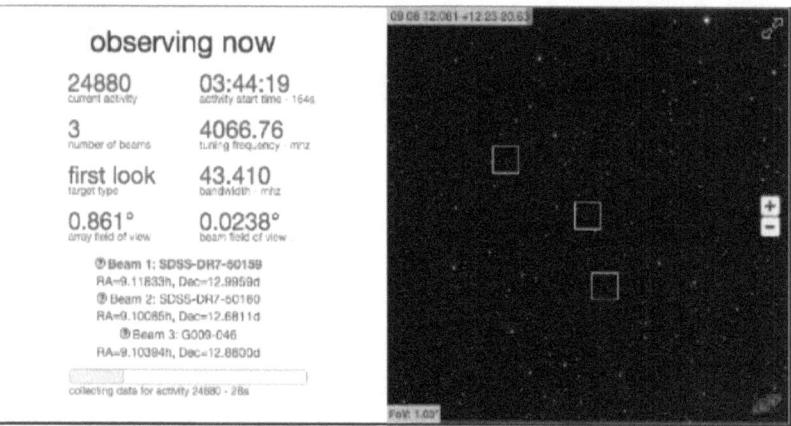

Screen capture from https://setiquest.info showing 3 beams under observation.

The analog voltage signals measured from the antenna are mixed (demodulated) from the GHz range down to lower frequencies and then digitized. The output of this processing is a stream of complex-valued time-series data across a range of frequency bandwidths of interest. At any given moment, the ATA can observe 108 MHz of spectrum within the 1 to 10 GHz range.

The software that controls the data acquisition system, analyzes the time-series data in real-time, directs repeated observations, and writes data out to disk is called SonATA (**SETI on the ATA**).

To find signals, the SonATA software calculates the signal power as a function of both frequency and time. It then searches for signals with power greater than the average noise power that persist for more than a few seconds. The representation of the power as a function of frequency and time are called spectrograms, or "waterfall plots" in the parlance of the field. To compute a spectrogram, a long complex-valued time-series data stream is chunked into multiple samples of about one-second worth of data. For each of these one-second samples, signal processing is applied (Hann windowing) and the power spectrum is calculated. Then, the power spectrum for each one-second sampled are ordered next to each other to produce the spectrogram. This is explained in pictures in a talk I gave earlier this spring (see slides 7–13).

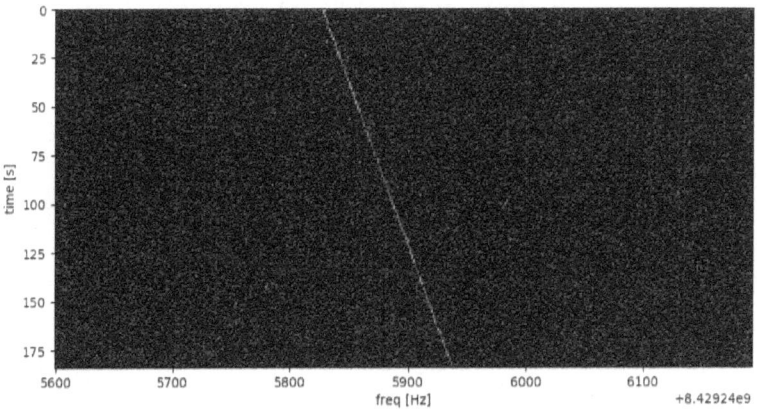

Signal observed at the Allen Telescope Array from the Cassini satellite while orbiting Saturn on September 3, 2014.

The figure above is an example of a classic "narrowband" signal, which is what SonATA primarily searches for in the data. The power of the signal is represented on a black & white scale. You can clearly see a signal starting at about 8.429245830 GHz and drifting up to 8.429245940 GHz over the ~175 second observation. Narrowband signals that have a large amount of power at a specific frequency (and hence, they have a "narrow" bandwidth). The reason that SonATA searches for these signals is because this is the kind of signal we use to communicate with our satellites, and it's how we suspect an E.T. civilization might transmit a signal to us if they were trying to get our attention. The central ("carrier") frequency of a narrowband signal, however, is not constant. Due to the rotation of the Earth and to the acceleration of the source, the frequency of the received signal drifts as a function of time, called Doppler Drift (not to be confused with Doppler Shift, though they are related).

The SonATA system was constructed to search primarily for narrowband signals. SonATA may label a signal as a "Candidate" when those narrowband characteristics are observed, the signal does not appear to have originated from a local source and is not found in a database containing known RFI signals. After a signal has been labeled as a Candidate, a new set of observations are made to test if that signal is persistent.

A persistent signal is one of the most important characteristics of a potential ET signal. First, SonATA tests to make sure it doesn't see the same Candidate signal in the other two beams (which would indicate RFI). It then forms a beam at a different point in the sky to ensure that it *doesn't* see the signal elsewhere. Then it looks back again to the same location. If it finds a signal *again,* the process is repeated. Each step along the way, the observed signal is recorded to disk in small files in an 8.5 kHz bandwidth about the frequency of the observation (as opposed to saving the entire stream of data over the full 108 MHz bandwidth). This pattern of observation can repeat up to five times, at which point the system places a phone call to a SETI researcher! (This has only happened once or twice in the past few years at the SETI Institute's ATA, I'm told.)

While SonATA is trained to find narrowband signals, it will often trigger on other types of signals as well, especially if there is a large power spike. There are many different "classes" of signals with a range of characteristics, such as smoothly varying drift rates, stochastically varying drift rates and various amplitude modulations. Additionally, these characteristics vary in intensity (they can be pronounced) in such a way that, overall, the different classes are not entirely distinguishable. Of course, this makes it hard to group and classify many of the real types of signals that are observed in SETI searches.

Clustering and classifying real SETI data

In 2015, the IBM Emerging Technologies start group joined up with researchers from the SETI Institute, NASA, and Swinburne University, forming this collaboration. The goal was two-fold: exercise some of IBM's new data management (Object Storage) and analytics (Apache Spark) product offerings to gain feedback, while providing significant computational infrastructure for SETI and NASA to explore the SETI raw data set. The 2013–2015 data set from the SETI Institute, which contains over 100 million Candidate and RFI observations and is a few TB in size, was transferred to IBM Object Storage instances. The Object Storage instances are located within the same data center as an IBM Enterprise Spark Cluster that was provisioned specifically for this collaboration. This computational setup has allowed researchers to spin through the data set many times over, searching for patterns in the observations. This data set is publicly available to citizen scientists via the SETI@IBMCloud project.

Over the following year, multiple attempts were made to cluster and classify the subset of Candidate signals found in the full data set. Some approaches were found to be more robust than others, but none were quite satisfactory enough for SETI Institute scientists to employ those techniques on a regular basis as part of their standard observational program.

Simulated signals and their classifiers

Due to the challenge of clustering and classifying the real SETI Candidate data, we decided to build a set of simulated signals that we could control and label. With a labeled set of data, we, or others, could train models for classification.

Based on manual observation, there are several classes of signals that SETI Institute researchers often observe. For this work, we decided to focus on just six of the different classes, plus a noise class. The signal classes were labeled 'brightpixel', 'narrowband', 'narrowbanddrd', 'noise', 'squarepulsednarrowband', 'squiggle', and 'squigglesquarepulsednarrowband'. The class names are descriptive of their appearance in a spectrogram.

All simulations were a sum of a signal and a noise background. They are described in detail below in order of increasing complexity. Be aware that all simulations were done entirely in the time-domain. The output data files were complex-valued time-series. All noise backgrounds were randomly sampled gaussian white noise with a mean of zero and RMS width of 13.0 for both the real and imaginary component. The spectrogram in the figures below were produced from a few example simulations. Also, the formula displayed in the figures do not fully characterize the simulations, but they are qualitatively useful for discussion.

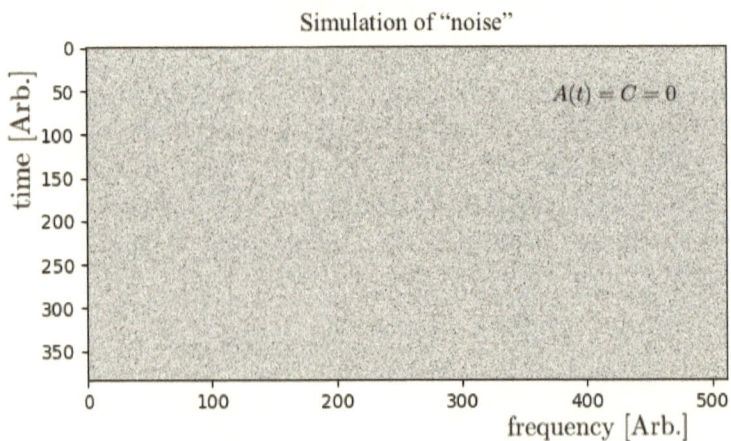

Gaussian white-noise with no signal.

Noise

The simulations labeled as 'noise', contained no signal, $A(t)=0$, plus the gaussian white noise background. In the full data set, there were 20k "noise" simulations.

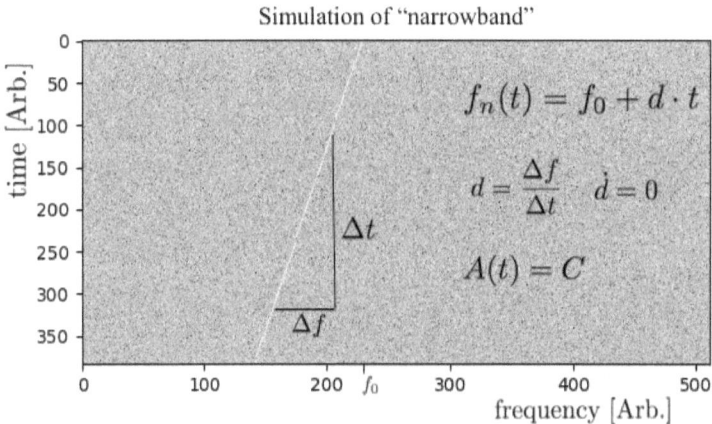

Typical narrowband signal with drifting central frequency.

Narrowband

Narrowband signals begin at some initial frequency, f_0, then change over time with a constant drift rate, d. Frequency drift indicates a non-zero acceleration between the transmitter and receiver. The amplitudes of these signals are constant throughout the simulation, $A(t) = C$. We simulated 20k narrowband signals, each one with a randomly selected initial frequency, f_0, drift rate, d, and signal amplitudes, C.

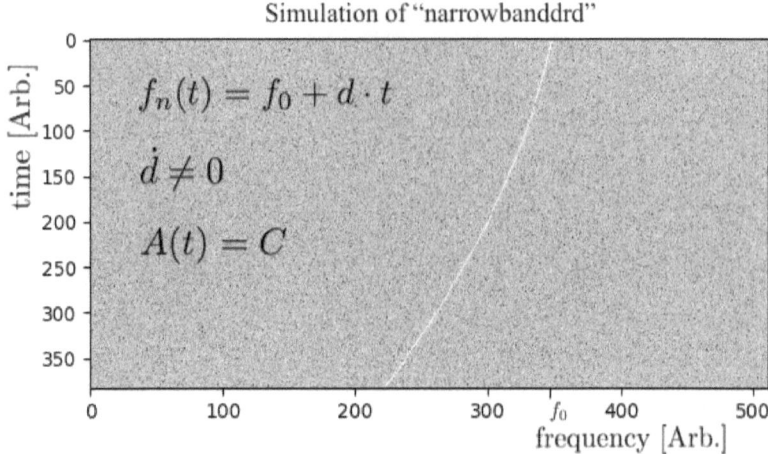

Narrowband DRD

Sometimes, signals are observed at the ATA where the drift rate does not remain constant. The frequency of the signal not only shifts in time, but shifts with an increasing or decreasing rate, as seen in the figure. These are labeled "narrowbanddrd", where DRD stands for "drift rate derivative". We simulated 20k narrowbanddrd signals, each one with a randomly selected initial frequency, f_0, drift rate, d, drift rate derivative, "d-dot", and signal amplitude, C.

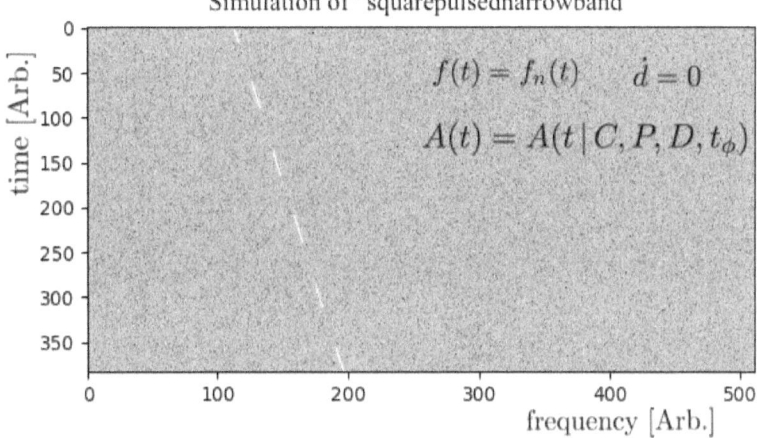

SquarePulsedNarrowBand

Another phenomenon observed in ATA data are narrowband signals that appear to have a square-wave amplitude modulation. The square-wave amplitude modulation, $A(t)$, is parameterized by its periodicity, P, duty cycle, D, and initial start time t_phi. Again, we simulated 20k signals of this type. The six variables that characterize these signals, fo, d, C, P, D and t_phi, were randomly chosen for each simulated signal.

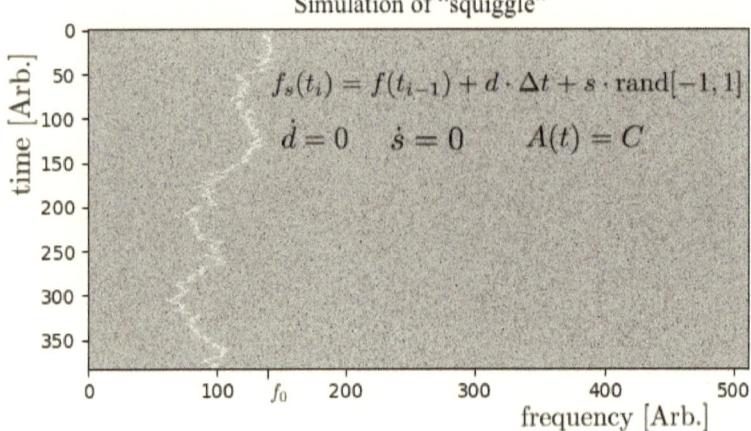

Squiggles

Signals with stochastically-varying frequencies often show up in ATA data, and are known as 'squiggles'. These signals were simulated by assigning an amplitude, s, to a randomly sampled value between -1 and 1. This simulates the random-walk of the signal's frequency as observed in the data. Note that the equation for the frequency as a function of time is slightly different here to describe the randomly shifting frequency. We simulated 20k squiggles with randomly chosen values for f_0, d, C and s.

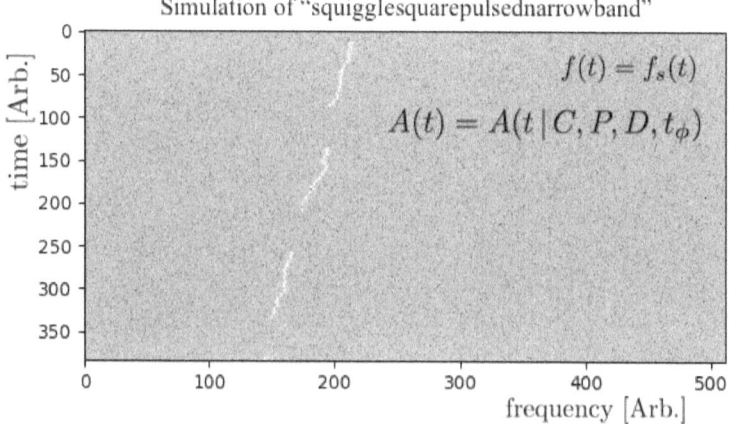

SquiggleSquarePulsedNarrowBand

We added a square-wave amplitude modulation to the squiggle signals in the same way was applied to the narrowband. We simulated 20k squiggles with randomly chosen values for *fo, d, C, s, P, D* and *t_phi*. (The title of this signal is a bit inconsistent in structure with the others because it contains the word "narrowband". A more appropriate signal name would have been SquarePulsedSquiggle.)

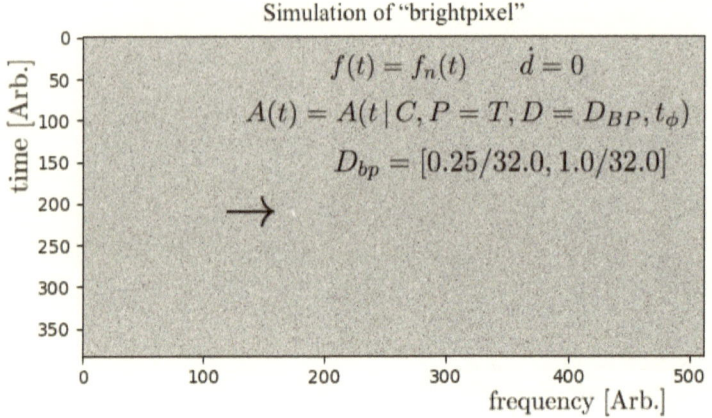

BrightPixels

Finally, signals called "brightpixels" were simulated. These are small blips of a signal where significant power is found for just a very short time at a specific frequency. In the real data at the ATA, however, these signals sometimes have broader spectrum. These are simulated in the exact same way as "squarepulsednarrowand", but with restricted range of values for the parameters that control the square-wave modulation. The periodicity, P, is fixed to the total length of the simulation, T. The duty cycle, D, is restricted to a very narrow range, which lets the bright pixel vary in size slightly. The initial offset time, t_phi, has the same full range as before. For the 20k simulated brightpixel signals, the fo, d, and C parameters were randomly selected, as before.

Distinguishability of signals: Range of simulation parameter values

The simulation parameters were confined to a range of values to create a simulated data set that closely represents the range of values found in real signals observed at the ATA. The amplitude of the background noise in all simulations was fixed by a gaussian distribution with zero mean and standard deviation width of 13.0 for both the real and imaginary components. This amplitude matches the typical noise amplitude observed at the ATA. All non-noise signals were simulated with an amplitude equal to a fraction of this noise amplitude for both the real and imaginary components.

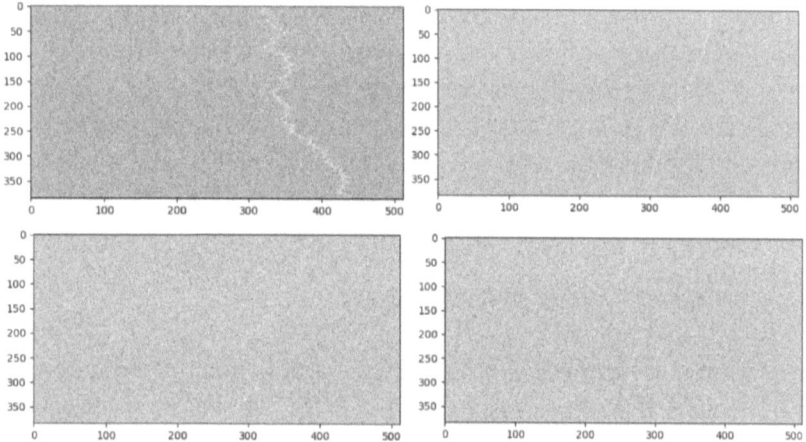

Four 'squiggle' simulations with different values for the squiggle parameter, s, and different SNRs. The simulation at the bottom right is not visible by the human eye.

The amplitudes, C, were uniformly distributed between 0.65 to 6.5 for most signals—brightpixel amplitudes ranged slightly higher, up to 9.75. Respectively, the range of signal to noise ratios, SNR, were, [0.05, 0.5] and [0.1, 0.75]. Note that these are signal and background amplitude values in the *time-domain*, and not in the spectrograms.

Parameters that controlled characteristics of the signals were uniformly distributed from nearly zero to values that produced a maximum effect like that observed in the real data. For example, the squiggle parameter, s, was distributed in the range [0.0001, 0.005]. As s approaches zero, squiggles begin to resemble narrowband signals. Similarly, the drift rate derivate value was evenly distributed in the range of [0.01, 0.08]. As the derivative approaches zero, these signals become indistinguishable from narrowband signals. In this case, we purposefully kept the lower-bound significantly above zero to keep this class of signal more distinguishable from narrowband.

For the square-wave amplitude modulation, the periodicity, P, was uniformly distributed from 15.6% to 46.9% of the total simulation time, T. The duty cycle, D, which controls the width of the square-wave, was uniformly distributed from 15% to 80% of the chosen periodicity, P. To simulate brightpixels, we used square-wave amplitude modulation with a fixed periodicity, $P=T$, and a very restricted duty cycle, D=[0.78%, 3.125%].

Simulation software & infrastructure

Simulation software was written in Java and Scala and executed on a 30-executor IBM Enterprise Spark cluster. Data were written to IBM Object Storage and IBM Db2 (formerly dashDB), both located within the same SoftLayer datacenter. There is no recorded simulation performance data, but anecdotally, about 1000 simulations could be created per minute, with the primary bottleneck being I/O to Object Storage and Db2. The software we used to simulate the SETI signals is still in a private repository. However, soon we will apply an Apache 2.0 License and release that code for those who are interested.

Training and test set details

In total, 140k signals were simulated and available for training classification models. Each simulated signal was placed in an individual file. Each file contained a JSON header, followed by raw bytes for the complex-valued time-series data. The bimester

Python package, which may be used to read and analyze real data from the ATA, was extended to read these simulation data files, facilitate signal processing and produce spectrogram. In the training data, the JSON headers contained the signal classification value and a UUID, whereas the JSON headers for the test data only contained a UUID. The UUIDs were used for reporting a team's test scores.

Two test sets were available for teams to score their trained models. The first test set, which we called the "preview" test set, allowed teams to score their models publicly. The second test set, called the "final" test set, was used for the final scoring and judging of classification models.

Each test set contained about 2400 simulated signals. However, the exact number of simulated signals for each class in the test sets were different. There were approximately 350 +- 50 simulated signals of each class. An unequal number of samples per class prevented attempts at artificially improving a team's score. If there was an equal number of samples per class, and teams became aware of this, that constraint could be exploited to modify class estimators and boost scores.

Teams were asked to build a .csv file scorecard. Each row of the scorecard file contained the UUID of the simulated file in the first position, along with seven numerical values that represented their model's degree of belief or probability for each class. The order of the values in each row were required to follow the alphabetical ordering of the class labels: brightpixel, narrowband, narrowbanddrd, noise, squarepulsednarrowband, squiggle, squigglesquarepulsednarrowband. For example, the line below indicates that a model scored the simulation test file "dbe38b359e70efb1a5fc2ea7bc4c619c", with a 99.997% probability of being a brightpixel.

Teams then submitted their scorecard for either the Preview or Final test set to the respective online scoreboard. Teams were allowed six submissions to the Preview Scoreboard, which allowed models to be updated and compared with other participants. However, only one submission was allowed to the Final Scoreboard. The scoreboards calculated the multinomial logistic regression loss (Log Loss) for the scorecard, which was the team's score. The team with the lowest Log Loss value was declared the winner.

The winning teams and results

All participants of the code challenge produced excellent results. Overall, they were much better than expected. The top teams were able to detect and identify signals that were buried deep into the noise.

The winning team, 'Effsubsee' (F_c), is Stéphane Egly, Sagar Vinodababu and Jeffrey Voien. They posted a classification accuracy of 94.99%! The second-place team was, 'Signet', who is Benjamin Bastian. He posted a classification accuracy of 94.67%. These teams differed only in their classification of a handful of the test cases.

Below are the classification accuracies and LogLoss scores for their models with the **preview test set** (scores for the final test set won't be published). In addition, an accompanying confusion matrix for each team's preview test set scorecard can be found in a Jupyter notebook in the ML4SETI repository.

```
                              precision    recall  f1-score   support

                 brightpixel    0.99262   0.84062   0.91032       320
                   narrowband    0.98592   0.92105   0.95238       380
                 narrowbanddrd    0.94693   0.97695   0.96170       347
                        noise    0.79433   1.00000   0.88538       336
        squarepulsednarrowband    0.96923   0.92375   0.94595       341
                      squiggle    0.99178   0.98638   0.98907       367
squigglesquarepulsednarrowband    0.98738   0.96904   0.97813       323

                   avg / total    0.95326   0.94615   0.94693      2414

Classification accuracy: 0.946147
Log Loss: 0.188138
```

Effsubsee's precision, recall and f1 scores for the ML4SETI Preview Test Set. Classification accuracy is equal to the average recall score.

```
                              precision    recall  f1-score   support

                 brightpixel    0.98571   0.86250   0.92000       320
                   narrowband    0.98864   0.91579   0.95082       380
                 narrowbanddrd    0.98529   0.96542   0.97525       347
                        noise    0.78824   0.99702   0.88042       336
        squarepulsednarrowband    0.94030   0.92375   0.93195       341
                      squiggle    0.99452   0.98910   0.99180       367
squigglesquarepulsednarrowband    0.99369   0.97523   0.98438       323

                   avg / total    0.95462   0.94739   0.94850      2414

Classification accuracy: 0.947390
Log Loss: 0.226332
```

Signet's precision, recall and f1 scores for the ML4SETI Preview Test Set. Classification accuracy is equal to the average recall score.

Interestingly, you'll notice, Effsubsee's LogLoss score for the preview test set was lower than Signet's score. However, Signet's classification accuracy was slightly greater.

Following Effsubsee and Signet, were Snb1 (Gerry Zhang) with 87.5% classification accuracy and LogLoss of 0.38467, Signy McSigface (Kevin Dela Rosa and Gabriel Parent) with 83.9% classification accuracy and LogLoss of 0.46575, and NulliusInVerbans with 82.3% classification accuracy and LogLoss of 0.56032. Their LogLoss scores are found on the Final Scoreboard.

First place and runner-up classification models

The Effsubsee and Signet teams have provided documentation and released their models under the Apache 2.0 license on GitHub.

Our approach was to experiment with various leading image classification architectures, and systematically determine the architecture that works best for the SETI signal data. We split the data into 5 parts, or "folds", with equal class distributions. Each model was trained on 4 folds, and the accuracy against the 5th fold was measured. (This is called the validation accuracy.) Below are the architectures that were constructed and the best validation accuracies we achieved for each class of architecture.

Residual Networks with 18, 50, 101, 152, 203 layers. The best model was the ResNet-101, with a single-fold validation accuracy of 94.99%.

Wide Residual Networks with 34x2, 16x8, 28x10 layers(x)expansion-factors. The best model was the **WideResNet-34x2**, with a single-fold validation accuracy of 95.77%.

Dense Networks with 161, 201 layers. The best model was the DenseNet-201, with a single-fold validation accuracy of 94.80%.

Dual Path Networks with 92, 98, 131 layers. The best model was the DPN-92, with a single-fold validation accuracy of 95.08%.

With very deep architectures, a common problem is overfitting to the training data. This means that the network will learn very fine patterns in the training data that may not exist in real-world (or test) data. While each of the five single-fold WideResNet-34x2 models had the highest validation accuracies, it was slightly overfitting to the training data. In contrast, a single-fold ResNet-101 performed the best on the preview test set, outperforming each of the other single-fold models. (This also makes the single-fold ResNet-101 an attractive candidate in a scenario where there are significant time constraints for prediction.)

However, for the winning entry, we used an averaged ensemble of five Wide Residual Networks, trained on different sets of 4(/5) folds, each with a depth of 34 (convolutional layers) and a widening factor of 2; the WideResNet-34x2.

To avoid overfitting, we combined the five single-fold WideResNet-34x2 in such a way that it takes a majority vote between them and eliminates inconsistencies. This was accomplished by a simple average the five results. As a result, the log-loss score for the five-fold WideResNet-34x2 was considerably better than the single-fold ResNet-101, with scores of 0.185 and 0.220, respectively.

In addition to their code, team Effsubsee placed the set of five model parameters in their GitHub repository. You can try the model yourself to calculate the class probabilities for a simulated signal, as demonstrated in this Jupyter notebook in IBM's Data Science Experience. (To use this notebook in your own DSX project, download the .ipynb file and create a new notebook from File.) Note that the Effsubsee original code was slightly modified to run their models on CPU. In general, with most modern deep learning libraries, this is relatively simple to achieve.

Second Place: Signet

Signet used a single Dense Convolutional Neural Net with 201 layers, as implemented in the torch vision module of porch. This was an architecture also explored by Effsubsee. It took approximately two days to train the model on Signet's GeForce GTX 1080 Ti GPU. Signet's code repository is found on GitHub.

Signet's model is also demonstrated calculating a simulated signal's class probabilities in a Jupyter notebook on IBM Data Science Experience. Some of Signet's code was slightly modified to run on CPU. (To use this notebook in your own DSX project, you can download the .ipynb file and create a new notebook from File.)

Run on GPU

Of course, you can also run these models locally or on a cloud server, such as those offered by IBM/SoftLayer or Nimbix Cloud, with or without a GPU. The setup instructions are rather simple, especially if you install Anaconda. But even without Anaconda, you can get away with pip installing almost everything you need. First, however, you will need to need to install CUDA 8.0 and should install cuDNN. After that, assuming you've installed Anaconda, it should be a handful of steps to get up and running.

Conclusions & next steps

The ML4SETI Code Challenge has resulted in two deep learning models with a demonstrated high signal classification accuracy. This is a promising first step in utilizing deep learning methods in SETI research and potentially other radio-astronomy experiments. Additionally, this project and the DSX notebooks above offer a clear picture of how a deep learning model, trained on GPUs, can then be deployed into production on CPUs when only inference about future new data need to be calculated.

The next most immediate task to be taken by the SETI/IBM team and the winning code challenge team, Effsubsee, will be to write an academic paper and to present this work at conferences. A future article will appear on arxiv.org and potentially in a suitable astro-physics journal.

Future technical updates

There are some improvements on this work that could be done to build more robust signal classification models.

New signal types & characteristics

There are two obvious advancements that can be made to train new deep learning models. First, more signal types can be added to the set of signals we simulate. For example, a sine-wave amplitude modulation could be applied to narrowband and squiggles, brightpixels could be broadened to include a wider range of frequencies, and amplitude modulation could be applied to narrowbanddrd. Second, the range of values for parameters that control the characteristics of the simulations could be changed. We could use smaller values for the squiggle parameter, and drift rate derivatives, for example. This would make some of the squiggle and narrowbanddrd signals appear very much like the narrowband signals. Obviously we expect classification models to become confused, or to identify those as narrowband more frequently as the parameters go to zero. However, it would be interesting to see the exact shape of the classification accuracy as a function of the amplitude of the parameters that control the simulations.

Different background model

We originally intended to use real data for the background noise. We observed the Sun over a 108 MHz bandwidth window and recorded the demodulated complex-valued time-series to disk. Overall there was an hour of continuous observation data. For the code challenge data sets, we used gaussian white noise, as described above. This was the version 3 (v3) data set. However, the version 2 data (v2) set **does** use the Sun observation as the background noise. The Sun noise significantly increases the challenge of building a signal classifier because the background noise is non-stationary and may contain random blips of signal of appreciable power.

The Sun noise could be used instead of gaussian white noise, along with the expanded ranges of signal characteristics in a future set of simulated data.

Object detection with multiple signals

We would like to perform not just signal classification but be able to find multiple different classes of signals in a single observation. The real SETI data from the ATA often contains multiple signals, and it would be very helpful to identify as many of these signal classes as possible. To do this, we'd need to create a labeled data set specifically for the purpose for training object detection models. In principle, all the components in the simulation software exist already to build such a data set.

Signal characteristic measurements and prediction

A useful addition to deep learning models would be the ability to measure characteristics of the signal. The SonATA system can estimate a signal's overall power, starting frequency and drift rate. Could deep learning systems go beyond that, especially for signals that are not the standard narrowband, and measure quantities that represent the amount of squiggle, the average change in the drift rate, or parameters about the amplitude modulation? The simulation software would need to be significantly updated to build such a system. The simulation signals would also need to include, beside the class label, the signal amplitude, frequency, drift rate, squiggle amplitude, etc., for machine learning models to learn how to predict those quantities. One solution may even be to perform signal classification with deep learning, and then use a more standard physics approach and perform a maximum likelihood fit to the signal to extract those parameters.

ML4SETI Code Challenge reboot

Even though the code challenge is officially over, it's not too late to attain the code challenge simulation data and build your own model. We've left the data available in the same locations as before, and the Preview and Final test sets and scoreboards are still online. You can form a team (or work on your own) and submit a result for the foreseeable future while these data remain publicly available. Additionally, you can join the ML4SETI Slack team to ask questions from me, SETI researchers, the top code challenge teams, and other participants.

There are a few places to get started. First, it may be informative and inspiring to watch the Hackathon video recap. Second, you should visit the ML4SETI GitHub repository and read the Getting Started page, which will direct you to the data sets and basic introduction on how to read them and produce spectrogram. Finally, you could take the example code above from Effsubsee and Signet and iterate on their results. Let us know if you beat their scores!

The ML4SETI code challenge would not have happened without the hard work of many people. They are Rebecca McDonald, Gerry Harp, Jon Richards, and Jill Tarter from the SETI Institute; Graham Mackintosh, Francois Luus, Teri Chadbourne, and Patrick Titzler from IBM. Additionally, thanks to Indrajit Poddar, Saeed Aghabozorgi, Joseph Santarcangelo and Daniel Rudnitski for their help with the hackathon and building the scoreboards.

Chapter 7

Wernher Von Braun

After World War II, several highly regarded German scientists were brought to the United States under a program called "Operation Paper Clip." These scientists were the elite of their field, working in rocket research and development under the leadership of dictator Adolph Hitler.

The leader of this group was the legendary Dr. Wernher von Braun. Braun was known as one of the world's leading authorities on space travel, and his expertise in rocket science was instrumental in America landing a spacecraft on the Moon.

From 1937 to 1945, von Braun was the technical director of the Peenemunde rocket research center, where the V-2 rocket –which devastated England, was developed. He worked on guided missiles for the U.S. Army and was later director of NASA's Marshall Space Flight Center. He became a celebrity in the 1950s and early 1960s, as one of Walt Disney's experts on the "World of Tomorrow." In 1970, he became NASA's associate administrator.

"In Robert Trundle's book, "Is ET Here," he discusses Clark C.

McClelland's relationship with a number of German scientists who were assigned to the ABMA." (Army Ballistics Missile Agency), first and foremost among them was Von Braun. These same men would eventually become part of NASA.

According to McClelland, there were occasional meetings of the MFA. (Manned Flight Awareness) These meetings were held in Cocoa Beach, Florida, and during breaks, McClelland was able to talk privately with some of these scientists.

One night, McClelland and Von Braun left the Cocoa Beach Ramada Inn and met in a back patio, giving McClelland a chance to ask the scientist a couple of questions privately. McClelland told Von Braun that he was aware that the group that Von Braun had led was located not too far from the site of the alleged Roswell crash of 1947. The group was launching V-2 rockets at the White Sands Testing Range.

McClelland states that Von Braun's eyebrows raised when he asked him the following questions;
"Did the Roswell Incident in fact happen?"
"Was an alien craft recovered along with alien bodies?
"Did you have a chance to go to the crash site?"

Von Braun lit a cigarette, thought for a second, and then began to discuss the crash openly.

McClelland vowed not to discuss Von Braun's remarks with anyone. McClelland kept the vow, and only after Von Braun's death, did he put in writing the amazing facts that he heard that night.

Dr. Von Braun explained how he and unnamed associates had been taken to the crash site after the bulk of the military personnel had left the scene. They did a quick once over of the site, Van Braun stated. He related how the exterior of the space craft was not metal as we know it, but appeared to be made of something biological, like skin.

McClelland's only thought at the time was that he was being told that the craft was made of something "alive."

And yes, there were alien bodies which were being kept in a medical tent near the UFO. The beings were small, very frail, and reptilian in nature. Von Braun compared their skin to rattlesnakes that he and his group had encountered at White Sands.

Von Braun was puzzled by the nature of the debris. The material was very thin, aluminum colored, like chewing gum wrapping, according to the scientist.

The description of the craft's interior was bizarre; it was very bare of instrumentation, as if the creatures and the craft were of a single unit.

McClelland relates that he left with his head buzzing from the shocking statements made by Von Braun. He found it difficult to not tell any of his colleagues about what he had been told. It was like finding a treasure and wanting to tell everyone you know about it. However, McClelland kept his vow to Von Braun, who died on June 15, 1977.

In 1947, a controversial event took place in New Mexico near the town of Roswell. The "Roswell Incident," as it has come to be known, remains the paramount case in UFO crash/retrieval history. In addition to the claims of a downed alien ship, alien bodies were said to have been recovered from the debris.

The USAF and Federal Government have kept a steadfast opinion that the object in question was a high-altitude balloon project, code named: Mogul. The project was designed to detect nuclear blasts in the USSR. The bodies that were recovered, according to the USAF, were parachute test dummies that had been released high above the desert and had eventually drifted into the "balloon" crash area. The USAF finally settled on this fabricated version of events and passed it off to the American public as truth.

During my long years of service in our national space program, I was very fortunate to come to know and exchange some very exciting data with former German scientists, who had been brought to the USA under Operation Paper Clip following WW II. These men were the elite of the German rocket programs controlled by Adolph Hitler. On many occasions I had the distinct privilege of speaking with Dr. Wernher von Braun, the leader of the elite group, and several other scientists who were assigned to the ABMA (Army Ballistics Missile Agency) launch crews at the Cape Canaveral launch sites. Eventually, these same men were incorporated into the new National Aeronautics and Space Administration (NASA) organization. During the periodic MFA (Manned Flight Awareness) meetings that were held at Cocoa Beach, I was able to talk freely and briefly with such scientists, particularly Dr. von Braun.

On one such occasion, he and I had taken a break and stepped out of the Cocoa Beach Ramada Inn into the back patio. I admitted that I was aware that he and his German Scientific team were located not too far from the crash site at that time. They were launching captured V-2 rockets from the White Sands Testing Range. On this night, I asked him a question concerning the Roswell Incident that caused his eyebrows to raise.

Did the Roswell Incident in fact happen, was an alien craft recovered along with alien bodies? Did you have a chance to go to the crash site?"

Dr von Braun was a cigarette smoker and he lit one up. He thought for a second, then proceeded to talk freely about his inspection of the crashed craft.

He trusted me to hear such astonishing events because I vowed to not report it to newspapers, magazines, television, etc. I never broke that vow. Since he is deceased, and the incident happened over fifty years ago, I am now disclosing what I heard. I have a right to speak about anything - even things that, according to certain agencies, "do not exist."

Dr. von Braun explained how he and his (unnamed, for now) associates had been taken to the crash site after most of the military were pulled back. They did a quick analysis of what they found. He told me the craft did not appear to be made of metal as we know metal on earth. He said it seemed to be created from something biological, like skin. I was lost as to what he indicated, other than thinking perhaps the craft was "alive." The recovered bodies were temporarily being kept in a nearby medical tent. They were small, very frail and had large heads. Their eyes were large. Their skin was grayish and reptilian in texture. He said it looked like the skin texture of rattle snakes he'd seen several times at White Sands. His inspection of the debris had even him puzzled: very thin, aluminum colored, like silvery chewing gum wrappers.

Very light and extremely strong. The interior of the craft was nearly bare of equipment, as if the creatures and craft were part of a single unit.

That's when I became lost in the moment. We returned to the awards ceremony, in which he participated, later bidding farewell. I went home with my head spinning from all I had heard. Keeping this quiet for many years was very difficult, especially with the temptations of having many friends and associates who believe in UFOs, ETs, etc. I never released this amazing data to Major Keyhoe and NICAP, or the public, until now. I considered my honor sacred when a vow was made.

This amazing interview with Dr. von Braun is only one of many events that I personally experienced as a space flight pioneer at Cape Canaveral and the Kennedy Space Center, Florida, from 1958 to 1992. (Anyone interested in assisting Mr. McClelland in the publication of his astounding book(s), please contact him at the address below his photo...)

Chapter 8

Stephen Hawking

Stephen William Hawking CH CBE FRS FRSA (8 January 1942 – 14 March 2018) was an English theoretical physicist, cosmologist, author, and director of research at the Centre for Theoretical Cosmology at the University of Cambridge. His scientific works included a collaboration with Roger Penrose on gravitational singularity theorems in the framework of general relativity and the theoretical prediction that black holes emit radiation, often called Hawking radiation. Hawking was the first to set out a theory of cosmology explained by a union of the general theory of relativity and quantum mechanics. He was a vigorous supporter of the many-worlds interpretation of quantum mechanics.

Hawking was a fellow of the Royal Society (FRS), a lifetime member of the Pontifical Academy of Sciences, and a recipient of the Presidential Medal of Freedom, the highest civilian award in the United States. In 2002, Hawking was ranked number 25 in the BBC's poll of the 100 Greatest Britons. He was the Lucasian Professor of Mathematics at the University of Cambridge between 1979 and 2009 and achieved commercial success with works of popular science in which he discusses his own theories and cosmology in general. His book *A Brief History of Time* appeared on the British *Sunday Times* best-seller list for a record-breaking 237 weeks.

Hawking had a rare early-onset slow-progressing form of motor neurone disease (also known as amyotrophic lateral sclerosis "ALS" or Lou Gehrig's disease) that gradually paralysed him over the decades. Even after the loss of his speech, he was still able to communicate through a speech-generating device, initially through use of a hand-held switch, and eventually by using a single cheek muscle. He died on 14 March 2018 at the age of 76.

Stephen Hawking on Alien Life, Extraterrestrials and the Possibility of UFOs Visiting Earth

Physicist Stephen Hawking died at his home in Cambridge, England, on Wednesday.

Hawking's earliest astrophysics work posited the existence of singularities, mathematically conforming black holes with Albert Einstein's general theory of relativity. Hawking established, along with Roger Penrose, the universe's origin as a singularity, i.e., a point in spacetime where traditional physical laws break down and gravity becomes infinite. His later work in quantum mechanics, inspired by collaboration with Soviet scientists Yakov Zel'dovich and Alexei Starobinsky, would mathematically indicate the finite entropy and evaporation of black holes as they emitted particles that came to be known as Hawking radiation. Though widely accepted as a breakthrough in theoretical physics, Hawking radiation and its resulting micro black holes have yet to be observed in experiments at CERN's Large Hadron Collider.

His work in theoretical astrophysics (and the 1988 publication of his bestselling book *A Brief History of Time*) made Hawking a celebrity—including appearances on *Star Trek: The Next Generation*, *The Simpsons* and *Futurama*—which allowed Hawking a prominent public platform for his beliefs outside of physics. An atheist, anti-war activist, BDS supporter and anti-capitalist, the overlap between Hawking's humanist politics and scientific interests found expression in his repeated public statements on the possibility of contact with extraterrestrial life.

Hawking took a conflicted position on alien life, at once promoting the search for extraterrestrial life and warning about the potential dangers of first contact with an alien species. His position on extraterrestrial life advocates two approaches: collecting intel and keeping as quiet as possible.

"There is no bigger question," Hawking said, while announcing his support for Breakthrough Listen, a $100 million program to search for alien communications via radio wave and visible light observations of 1 million nearby stars and 100 galactic centers. "It is time to commit to finding the answer, to search for life beyond Earth."

In 2010, Hawking worried what that answer would bring, describing the dangers of first contact with aliens in a Discovery Channel documentary. "If aliens visit us, the outcome would be much as when Columbus landed in America, which didn't turn out well for the Native Americans," Hawking says. "We only have to look at ourselves to see how intelligent life might develop into something we wouldn't want to meet."

"Such advanced aliens would perhaps become nomads, looking to conquer and colonize whatever planets they can reach," Hawking said in the documentary, *Into the Universe with Stephen Hawking*.

Why Stephen Hawking and the world's top scientists say this massive object hurtling through space could be an alien spaceship

BRIT scientist Stephen Hawking is among the leading space experts today using top-of-the-range technology to scan a vast cigar-shaped object for signs of alien life.

Thought to be an interstellar asteroid the object is about a quarter of a mile long, 260ft wide and hurtling through our solar system at 196,000mph.

PA:PRESS ASSOCIATION

Scientists believe that a V-shaped asteroid could be a spaceship

Stephen Hawking speaks on stage during a Breakthrough Starshot press conference at One World Observatory in New York City on April 12, 2016. Breakthrough Starshot is one of several projects by the Breakthrough Initiative, who also organized Breakthrough Listen. Bryan Bedder/Stringer/Getty Images

Absent knowledge about alien life, hawking urged documentary viewers to analogize their likely behavior to ours. Hawking noted that first encounters throughout our own history rarely begin with: "I'll pop the kettle on. Milk? Sugar?"

He would reiterate this theme in a later documentary. "One day, we might receive a signal from a planet like this," he says in *Stephen Hawking's Favorite Places* of the newly discovered world of Gliese 832c. "But we should be wary of answering back."

During the announcement for Breakthrough Listen, Hawking said: "We don't know much about aliens, but we know about humans. If you look at history, contact between humans and less intelligent organisms have often been disastrous from their point of view, and encounters between civilizations with advanced versus primitive technologies have gone badly for the less advanced. A civilization reading one of our messages could be billions of years ahead of us. If so, they will be vastly more powerful, and may not see us as any more valuable than we see bacteria."

While Hawking expresses near certainty that alien life exists in the universe, he does not believe aliens have visited Earth in UFOs or at any point in history. "Why hasn't the Earth been visited, and even colonized?" Hawking wrote on his official website. "I discount suggestions that UFOs contain beings from outer space. I think any visits by aliens, would be much more obvious, and probably also, much more unpleasant."

In the essay Hawking describes some of the possibilities for the universe's seeming silence, speculating that intelligence may be one of many possible evolutionary outcomes or, mostly darkly of all, the possibility that "intelligent life destroys itself."

"I very much hope it isn't true," Hawking wrote

The team, led by Hawking, is using high tech scanners to investigate the object dubbed "Oumuamua".

The team, named Breakthrough Listen are using the world's largest maneuverable radio telescope at Green Bank, West Virginia, to track the celestial object for ten hours.

The boffins are listening for electromagnetic signals which cannot be produced in nature.

If found that would be the biggest indication ever that extraterrestrial intelligence could be involved.

Stephen Hawking on time travel, M-theory, and extra-terrestrial life

Invite time travelers to a party late? "I sat there a long time, no one came."

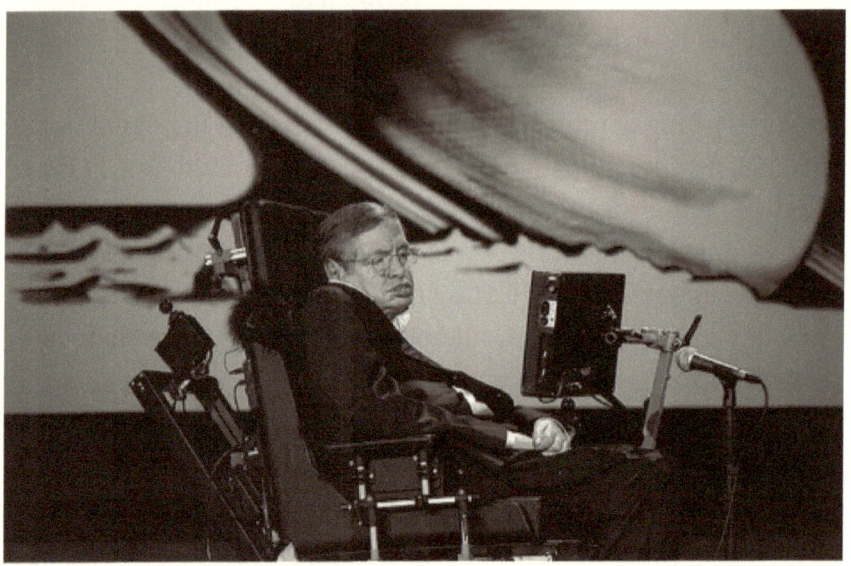

Stephen Hawking at a NASA fiftieth anniversary event in 2008.

Share this story

Stephen Hawking is near-universally recognized as a brilliant scientist, and one small facet of his complex genius is the Rockstar ability to democratize scientific knowledge. It's becoming increasingly important as purveyors of pop culture continue to eschew acknowledgment and respect for the most basic scientific principles. Hollywood in particular has continued its marginalization of scientific knowledge with blatant disregard in recent films.

But Hawking acts like a great counter force against anti-intellectual movements. He takes complex scientific principles and explains them, so the public can understand and, more importantly, appreciate the science behind them. He inspires people to want to know more about Calabi–Yau manifolds and multiverses. That is why Stephen Hawking rocks our scientific world.

So, I (and everyone around me) felt a consummate air of curiosity in the Paramount Theater lobby on June 16, as Dr. Hawking arrived for an appearance at the Seattle Science Festival. But what was everyone curious about? Was it Dr. Hawking's motor neuron disease? Or how his speech synthesizer works, perhaps?

The following are the press questions from Seattle-area colleagues that I read to Dr. Hawking at a press event before his main program. The 70-year-old scientist keynoted a Saturday night symposium that also featured Leroy Hood and Jack Horner, with Hawking discussing his life and Nobel aspirations among other topics.

What would it take to make time travel a reality, and how would that affect the present reality? (via Arik Korman, 95.7 KJR)

We are all travelling forward in time anyway. We can fast forward by going off in a rocket at high speed and return to find everyone on Earth much older or dead. Einstein's general theory of relativity seems to offer the possibility that we could warp space-time so much that we could travel back in time. However, it is likely that warping would trigger a bolt of radiation that would destroy the spaceship and maybe the space-time itself. I have experimental evidence that time travel is not possible. I gave a party for time-travelers, but I didn't send out the invitations until after the party. I sat there a long time, but no one came.

Horner and Hood at the symposium

Jack Horner entertained the audience by waxing about his work at Montana State University, where his goal is to use the tools of evolution to recreate the dinosaurs. The gist of the effort: his Montana labs are attempting what is known as "reverse evolution," creating a dinosaur from existing bird species such as the chicken. Horner points out that all life is interrelated, as species have long descended from one another with certain modifications.

So, Horner's Montana lab is tweaking evolution's dialectic via complex procedures on embryos and making probes with extracted DNA. They then replace the probe to control the growth of "experimental atavisms"—long suppressed evolutionary dinosaur characteristics such as a long tail, three-fingered claws, and dinosaur teeth. Horner's goal is to grow the world's first "chickasaur." And he points out it's best to start out small because it's easier to catch if something goes wrong.

Leroy Hood's presentation touched on what he calls P4 medicine: predictive, preventative, personalized, and participatory. It is predictive because future medicine will draw from what will be a completed data set of the human genome in about ten years and will make use of massive pools of digitized bio-information. "Preventative" because the approach to drugs will target discovery of illness via the use of such large repositories of digital genomic data. The P4 model should see the cost of medicine go down, allowing a focus on wellness from the continued mining of millions of data points for ongoing patient diagnoses. Each patient will have a unique genomic profile, so the approach can be tailored to each patient. Hood even imagines patient-driven social networks for individuals to share information on disease and wellness.

If M-theory is the only candidate for a complete theory of the universe, what's the best evidence that you think will be found to support the theory? Lacking that evidence, isn't M-theory just another kind of religion? (via Alan Boyle, MSNBC)

M-theory is the only theory that *seems* to have all the properties we would expect of a complete and consistent "history of everything," but that may just reflect our lack of imagination. If M-theory is correct, it predicts that every particle should have a super-partner. So far, we have not observed any super-partners, but the hope is that they will be found at the LHC. If they are discovered, it will be strong evidence for M-theory. On the other hand, it they are shown not to exist, that will be exciting, because we will learn something new.

How would you describe your quality of life? What do you miss most from before the onset of ALS? (via Arik Korman, 95.7 KJR)

Although I am severely disabled and on a ventilator, my quality of life is pretty good. I have been very successful in my scientific work and have become one of the best-known scientists in the world. I have three children and three grandchildren so far. I travel widely, have been to Antarctica, and have met the presidents of Korea, China, India, Ireland, Chile, and the United States. I have been down in a submarine and up on a zero-gravity flight, in preparation for the flight into space that I'm hoping to make on Virgin Galactic. Despite my disability, I have managed to do most things I want. My main regret is that it has prevented me from playing with my children and grandchildren as fully as I want.

John Gribben recently argued that we are almost certainly the only intelligent life in the Milky Way—do you think he's right or wrong, and why? Also, Seth Shostak argues that even if there are other intelligent civilizations out there, it's too late for us to keep quiet about our existence, because it's possible to pick up the signals we've sent out over the past seventy years. So, isn't it too late for us to keep quiet, and shouldn't we be thinking about upgrading our defenses against the alien hordes? (via Alan Boyle, MSNBC)

We think that life develops spontaneously on Earth, so it must be possible for life to develop on suitable planets elsewhere in the universe. But we don't know the probability that a planet develops life. If it is very low, we may be the only intelligent life in the galaxy. Another frightening possibility is intelligent life is not only common, but that it destroys itself when it reaches a stage of advanced technology.

Evidence that intelligent life is very short-lived is that we don't seem to have been visited by extra-terrestrials. I'm discounting claims that UFOs contain aliens. Why would they appear only to cranks and weirdos? Do I believe that there is some government conspiracy to conceal the evidence and keep for themselves the advanced technology the aliens have? If that were the case, they aren't making much use of it. Further evidence that there isn't any intelligent life within a few hundred light years comes from the fact that SETI, the Search for Extra Terrestrial Life, hasn't picked up their television quiz shows. It is true that we advertise our presence by our broadcast. But given that we haven't been visited for four billion years, it isn't likely that aliens will come any time soon.

A team of scientists lead by Stephen Hawking are scanning the object for signs of life

Hawking and his colleagues at Breakthrough Listen said: "Researchers working on long-distance space transportation have previously suggested that a cigar or needle shape is the most likely architecture for an interstellar spacecraft, since this would minimize friction and damage from interstellar gas and dust."

Other anomalies from normal asteroids are its long slender shape and flight path.

Experts currently say that it is made of something dense, most likely rock but possibly metal.

Astronomers from the University of Hawaii spotted Oumuamua in October passing the Earth.

The space rock is currently twice as far from us as the sun and will eventually shoot passed Jupiter next year.

It is the first object discovered in the solar system that appears to have originated from another part of the galaxy.

Although thought to be an asteroid, Oumuamua's elongated cigar shape hundreds of meters in length but only one tenth as wide is highly unusual for a typical space rock.

Travelling at up to 196,000mph, the object's high speed also suggests that it is not gravitationally bound to the sun but is destined to head back out of the solar system.

Stephen Hawking launches biggest-ever search for alien life

British cosmologist Stephen Hawking has launched the biggest-ever search for intelligent extraterrestrial life in a 10-year, $135 million project to scan the heavens.

Russian Silicon Valley entrepreneur Yuri Milner, who is funding the Breakthrough Listen initiative, said it would be the most intensive scientific search ever undertaken for signs of extraterrestrial intelligent life.

"In an infinite universe, there must be other occurrences of life," Mr. Hawking said at the launch event at the Royal Society science academy in London.

"Somewhere in the cosmos, perhaps, intelligent life may be watching.

"It's time to commit to finding the answer, to search for life beyond Earth. We must know."

The project will use some of the biggest telescopes on Earth, searching far deeper into the universe than before for radio spectrum and laser signals.

"We are launching the most comprehensive search program ever," Mr. Milner said.

"Breakthrough Listen takes the search for intelligent life in the universe to a completely new level."

Mr. Milner said the scan would collect more data in one day than a year of any previous search, tracking the million closest stars, the center of the Milky Way and the 100 closest galaxies.

It's a huge gamble, of course, but the pay-off would be so colossal ... even if the chance of success is small.

British astronomer Martin Rees

"We should not read too much into the lack of confirmed signals," said the former physics student, who is named after Yuri Gagarin, the first man in outer space.

Earth's telescopes would be able to detect a signal from similarly-advanced technology sent from the center of the Milky Way.

A signal from Andromeda, the nearest major galaxy, would need only the power of two times the Three Gorges Dam in China to reach Earth.

"We don't need to assume that civilization is way more developed than we are," Mr. Milner said.

Parkes Observatory in NSW to play vital role

Australia will play a crucial role in the project, with the Parkes Observatory in New South Wales signing a multi-million-dollar contract to scan radio waves for life in the cosmos.

We know there's lots of worlds out there, but whether there's little green men on there is something we still don't know.

Professor Matthew Bailes, Swinburne University

The CSIRO will contract a quarter of the telescope's time for five years to scan for potential radio signals from galaxies beyond our reach.

Professor Matthew Bailes, from Melbourne's Swinburne University, will be the project's lead investigator at the Parkes Telescope.

"Radio waves are a very efficient way of transmitting information and it's likely that aliens, if they're into interstellar communication at all, would be using the radio part of the spectrum," he said.

Who is Russian billionaire Yuri Milner?

Internet entrepreneur Yuri Milner explains his interest in science by saying he was a physicist in a past life.

"We know there's lots of worlds out there, but whether or not there's little green men on them is something we still don't know."

Cosmologist Professor Paul Davies from Arizona State University has spent his career studying the question of "what else is out there?" and is also excited by the project.

But he said there would be several big questions to ask if the project did find intelligent life.

"Should we respond? Who gets to speak for us? What do we do next?" he said.

"There are no easy answers to these but if we're going to engage in a dialogue with an alien civilization this is something that needs the consideration of everybody on the planet."

Possibility of finding aliens rises a billionfold

Martin Rees, Britain's astronomer royal and one of the project leaders, said modern technology allowed much more sensitive searches than ever before, though he cautioned against expectations of finding intelligent alien life.

The $100m question: are we alone?

"It's a huge gamble, of course, but the pay-off would be so colossal

... even if the chance of success is small," the astrophysicist said.

The possibility of finding life had however effectively risen a billionfold through the identification of billions of Earth-like planets in the Milky Way, he said.

"Is there life out there? We may not answer it, but this gives a bigger chance that it may be answered in our lifetime," he said.

The program will be 50 times more sensitive than previous searches, and cover 10 times more of the sky, experts said.

It will scan at least five times more of the radio spectrum, and 100 times faster, while in tandem undertake the deepest and broadest-ever search for optical laser transmissions.

The initiative was launched on the 46th anniversary of the first manned Moon landing.

It is allied with the Breakthrough Message project, an international competition to create digital messages that represent humanity.

There is no commitment to send any messages into space, and the project should spark discussion about whether humans should be sending messages at all out into the void.

PA:PRESS ASSOCIATION

A team of scientists lead by Stephen Hawking are scanning the object for signs of life

Hawking and his colleagues at Breakthrough Listen said: "Researchers working on long-distance space transportation have previously suggested that a cigar or needle shape is the most likely architecture for an interstellar spacecraft, since this would minimize friction and damage from interstellar gas and dust."

Other anomalies from normal asteroids are its long slender shape and flight path.

Experts currently say that it is made of something dense, most likely rock but possibly metal.

Astronomers from the University of Hawaii spotted Oumuamua in October passing the Earth.

The space rock is currently twice as far from us as the sun and will eventually shoot passed Jupiter next year.

It is the first object discovered in the solar system that appears to have originated from another part of the galaxy.

Although thought to be an asteroid, Oumuamua's elongated cigar shape hundreds of meters in length but only one tenth as wide is highly unusual for a typical space rock.

Travelling at up to 196,000mph, the object's high speed also suggests that it is not gravitationally bound to the sun but is destined to head back out of the solar system.

PA:PRESS ASSOCIATION

Researchers used the Green Bank radio telescope in Virginia, US, to get a better look at the space rock

A statement from the £75million Seti project Breakthrough Listen, launched by Russian digital tech mogul Yuri Milner in 2015, said: "Researchers working on long-distance space transportation have previously suggested that a cigar or needle shape is the most likely architecture for an interstellar spacecraft, since this would minimize friction and damage from interstellar gas and dust.

"While a natural origin is more likely, there is currently no consensus on what that origin might have been, and Breakthrough Listen is well positioned to explore the possibility that Oumuamua could be an artifact."

The Breakthrough Listen team is using the Green Bank radio telescope in West Virginia, US, to study Oumuamua, which is named after the Hawaiian term for "scout" or "messenger".

From 8pm UK time on Wednesday, December 13, the giant dish - the largest fully steerable radio telescope in the world - will "listen" to the object across four radio frequency bands spanning one to 12 gigahertz.

AP:ASSOCIATED PRESS

In a statement the researchers said that the architecture looks like that of a spaceship after it was seen from the Pan-STARRS1 Observatory in Hawaii

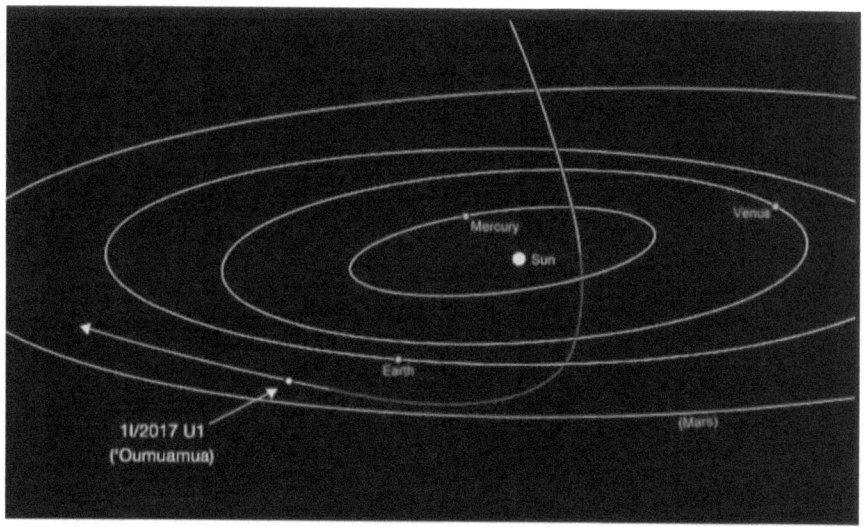

NASA

This diagram from NASA shows the flight path of the Oumuamua as it heads back out of the solar system

GETTY - CONTRIBUTOR

The research is being funded by California-based Russian billionaire Yuri Milner

Lead scientist Dr Andrew Siemion, director of the Berkeley Seti Research Centre in California, said: "Oumuamua's presence within our solar system affords Breakthrough Listen an opportunity to reach unprecedented sensitivities to possible artificial transmitters and demonstrate our ability to track nearby, fast-moving objects.

"Whether this object turns out to be artificial or natural, it's a great target for Listen."

The object is currently about two astronomical units (AU) from Earth, or twice the distance between the Earth and sun.

At this distance it would take less than a minute for the Green Bank telescope to detect an omnidirectional transmitter with the power of a mobile phone.

Chapter 9

Carl Sagon

""The Cosmos is all that is, or ever was, or ever will be. Our contemplations of the Cosmos stir us — there is a tingling in the spine, a catch in the voice, a faint sensation as if a distant memory, of falling from a great height. We know we are approaching the grandest of mysteries.

—Carl Sagan, *Cosmos*

Carl Sagan (1934–1996) was an American astronomer who did much to popularize science, especially astronomy, during his illustrious career. He co-wrote and presented *Cosmos: A Personal Voyage*, a television series that kicked ass. His legacy lies mostly in his advancement of humanism. He found a profound spirituality in experiencing the wonder and majesty of the universe.

He is known for remarking how people were made of "star stuff." And, by the by, he was right: H → He → Li → Be → B → C → N → O → Fe, and on through various nuclear processes that synthesize the elements up to iron (Fe) — and it takes a supernova (or a particle accelerator!) to make a heavier nucleus.

Sagan was also very interested in extraterrestrial life, including UFO sightings (though he was a skeptic on the matter); he heavily promoted the search for extraterrestrial intelligence.

In 1974 he was enlisted by the American Association for the Advancement of Science (AAAS) to lead the attack at their conference debunking the theories of Immanuel Velikovsky. His rather cavalier performance at this event was perhaps a salutary lesson in the fight against woo: far from stamping out the growing cult of Velikovsky, he arguably further stoked it, by allowing a few schoolboy errors to creep into his math and peppering his talk with cheeky jokes. This use of levity and lack of rigor was seized upon by the Velikovsky faithful as evidence that Thine Mainstream were not playing fair and were trying to suppress the truth.

A prominent non-believer, Carl Sagan is said by many fundamentalists to have converted to Christianity on his deathbed, the testimony of his widow notwithstanding. This is a cowardly attack because they know Dr. Sagan can't come back to make a refutation. Some of them point to a quote from his famous novel *Contact* about "an intelligence that antedates the Universe" as evidence of this change in beliefs, but fellow science fiction author Robert J. Sawyer, also an agnostic, has pointed out Sagan was no more obligated to believe in the religious aspects of that novel than *Star Wars* creator George Lucas was to believe in The Force.

A recent development is the celebration of "Carl Sagan Day," which is celebrated on his birthday, November 9. The first celebration of the event, in 2009, was attended by James Randi and Phil Plait as guest speakers.

He was and remains to this day arguably one of the greatest science communicators ever.

How Carl Sagan Worked

The famous scientist and popularizer of science poses in his Cornell lab circa 1974, turtleneck sweater and all.

Resting on the foreign, reddish surface of Mars, a compartment in NASA's Phoenix spacecraft carries a heartfelt message from one of the world's most fascinating researchers. It reads:

"Whatever the reason you're on Mars, I'm glad you're here, and I wish I was with you."

Crafting that message, one of a collection of DVD dispatches intended for future Mars settlers, was just another day for Carl Sagan. Did anyone delight in describing the universe, with its billions of galaxies, as much as Sagan? Who else would launch a record with Bach and Mozart into space?

His yearning to understand the cosmos wasn't limited to the expansive blackness beyond Earth's atmosphere. He believed everyone should share in the wonders of science and the universe -- a legacy that lives on through colleagues, fans and family to this day. Some present-day admirers mix Sagan's musings in music form ("A Glorious Dawn," anyone?). Humbly, Sagan reminded us that we too are made of "star stuff."

Perhaps the phrase "billions upon billions of stars" conjures up breath-taking images from the PBS series "Cosmos." Or you may be familiar with Sagan's skeptical attitude or search for extraterrestrial life. He also left his mark on political and religious discussions, recognizing that science affects many facets of life. Yet one Sagan biographer notes that the greatest dichotomy of Sagan's life was Sagan himself -- he was both prophet and skeptic; someone who teetered between being uplifted by creativity and grounded by the facts [source: Davidson].

Let's get to know the man whose idea of fun entailed characterizing the mysterious surfaces of planets, describing the chemistry required for life and pondering humans' place in the enormous universe. His legacy spans the space program, academia and popular media, and it all began in Brooklyn.

The Birth of a Star

Born to a Jewish-Ukrainian family in Brooklyn, New York, Sagan was the son of hardworking parents who did their best to give him opportunities and protect him from the grim realities of the Holocaust, which negatively affected extended family abroad. Sagan's mother had particularly high aspirations for her son. Some say her desire for Sagan's success was to make up for opportunities she lacked in life [source: Davidson].

Thoughts of unseen worlds started taking root in Sagan when the World's Fair set up in New York in 1939. As a young boy, he became intrigued by exhibits touting the role of science and technology in humanity's future. The displays hosted model skyscrapers and cities; booths enticed him with the new invention of the television. Sagan delighted in the future's potential -- it would be ripe with technology and science [source: Poundstone].

His inspirations drove him toward answering fundamental questions about the natural world -- a behavior routinely satisfied with his very own children's library card. Museums enraptured young Sagan, offering glimpses of other parts of the world and the exotic [source: Davidson].

Sagan received his bachelor's and master's degree in physics, and a doctorate in astronomy from the University of Chicago, all while he was still in his 20s. After that, he did a stint at the Smithsonian Astrophysical Observatory and taught at Harvard University. He soon found a position at Cornell University in Ithaca, N.Y., where he settled into a more permanent position as a professor and director of the Laboratory for Planetary Studies.

During and after his studies, Sagan worked with NASA, advising the agency on several space projects, including the Apollo, Voyager, Viking, Pioneer, Mariner and Galileo missions.

But Sagan didn't peak as a science star until he began working on the educational PBS show "Cosmos." Sagan's books and essays also catered to the masses, granting him numerous awards, including the prestigious Pulitzer Prize for his book "Dragons of Eden."

Sagan first wed biologist Lynn Margulis in 1957; then, artist Linda Salzman in 1968; and finally, author Ann Druyan in 1981. He was the father to five children, but his commitment to work sometimes took a toll on providing a normal family life [source: Poundstone].

What's in a Planet?

Like most astronomers, Carl Sagan relied on indirect measurements of faraway planets and galaxies for his research. Years later, even as more precise tools have emerged, his work on planetary atmospheres remains relevant. His passion also led to him creating the Planetary Society.

Take his description of Venus's atmosphere, for example. In the early 1960s, Sagan tackled why the second planet in our solar system trapped heat in its atmosphere. He hypothesized that a **greenhouse effect** kept Venus at a toasty 900 degrees Fahrenheit (500 degrees Celsius). As sunlight hits the surface of the planet, gases in its atmosphere trap the heat rather than let it escape. Sagan's conclusions opened the door to discovering a similar phenomenon on other planets -- even Earth.

By studying other planets' atmospheres and chemical cycles, Sagan discovered that many of the same processes were at play on our home planet. Despite Earth's early, icy history, geology shows that not all water was bound up in ice. But what propelled the planet to the warmer, wetter place it is today? That's what puzzled Sagan and his colleague George Mullen, who labeled the issue the **faint young sun paradox** in 1972. Since stars gain energy and luminosity as they age, the sun emitted more and more energy that could have helped Earth thaw. Yet Sagan and Mullen suggested that an increasingly powerful sun couldn't have been entirely responsible for melting young Earth; rather, other greenhouse gas cycles may have accelerated warming. Scientists are still trying to piece together the puzzle today, but Sagan helped draw attention to the paradox.

Mars enchanted Sagan, too, driving his strong desire to explore the red planet. He hammered out the logistics of space missions at the NASA Jet Propulsion Laboratory and described Mars' seasons, citing the planet's seasonal storms as causes behind its shifting features.

The famous astronomer also believed there was more to Saturn's moon Titan than meets the telescoped eye. He hypothesized that complex organic molecules lent a reddish hue to the distant moon, which ended up being true.

In Pursuit of (Other) Life

Are we alone in the universe? Can physics and math -- fundamental truths that underpin the universe -- act as symbolic messages to communicate with other life?

Carl Sagan wondered similar questions. Perhaps more than anyone, Sagan wanted to find other intelligence in the universe, but he was grounded by the lack of proof [source: Poundstone]. He famously coined the phrase, "Extraordinary claims require extraordinary proof," and at times dismissed his own sci-fi fantasies of aliens lurking around the cosmic corner. In a sense, Sagan's obsession with finding other life made him more relatable to the public. At the time, it was rare to see scientists comment on such things.

His 1985 novel "Contact," which was later made into a motion picture, portrayed one scientist's transmitting signals into space with the hope of intercepting other life. The novel certainly drew inspiration from Sagan's search for ET at the SETI Institute and from colleague Jill Tarter in particular. Sagan also strove to understand the chemical recipe necessary for basic life outside of Earth to take root, a discipline known as **exobiology**. He promoted using biology, chemistry and physics to probe life's origins.

But his most memorable space projects were less about hard science and more about humanity and love.

Should ET ever intercept a NASA spacecraft, Sagan wanted to be ready. This explains the golden record on Voyager 1 and 2, each containing "sounds" from the natural world and cultures, and 115 "images" representative of the diversity of life on Earth [source: NASA Jet Propulsion Laboratory].

Conclusion

NASA

NASA Admits It Is in Contact with Alien Species and Just Forgot to Mention It

A conspiracy site republished a satirical claim that NASA admitted alien contact but failed to disclose that information, presuming everyone already knew about it.

CLAIM

NASA admitted that they were in contact with aliens but failed to officially disclose that information, presuming everyone already knew about it. See Example (s)

EXAMPLES

Collected via e-mail, September 2016

A few articles have claimed that a NASA spokesperson admitted that there have been aliens visiting the Earth for thousands of years, and NASA assumed that everybody already knew that aliens were real because of all the science fiction about them.

ORIGIN

On 19 September 2016, the conspiracy web site Disclose.tv published an article reporting that NASA, the federal agency that oversees the U.S. civilian space program, had casually admitted they were in contact with aliens but had never formally announced that information because they believed everyone was already aware of it:

According to reports, Trish Chamberson, an official spokesperson from NASA has confirmed the existence of extra-terrestrial life and has claimed that aliens have been visiting planet Earth for thousands of years.

NASA SPOKESWOMAN CONFIRMS THE EXISTENCE OF ALIENS

During the two-hour briefing, Chamberson confirmed that several theories which had previously been dismissed as groundless speculation from fringe enthusiasts are grounded. Chamberson made various sensational allegations during the interview, claiming that the alien species known as the Greys have been visiting Earth for thousands of years and that they may have had a hand in the construction of megastructures such as the ancient pyramids of

Giza and various other buildings dotted around the world.

There are so many films, documentaries and TV programs on aliens, that we thought everyone was aware of them by now[.]

Chamberson went on to confirm various theories about alien mining operations in the solar system. She claimed one of the mines was on the far side of the Moon and that various planets in the solar system were being assessed for minerals. Recently, she claimed aliens have begun to mine Jupiter, which is why observers have been able to see several apparently new rings appearing around the gas giant.

Sorry. We just kind of assumed everyone knew about it[.]

Disclose.tv didn't provide any sources for the attention-grabbing claim, but it was easily traced back to an article published by satirical web site Waterford Whispers a month earlier:

"We do apologize for this mix up, the whole thing just slipped our mind," another scientist explained, "we were so busy back-engineering their technology, we simply forgot all about it. They even have a base on the far side of the moon and are currently mining several planets in our solar system for minerals. They've only started on Jupiter recently, hence the new rings around it. It's all good though, they're a nice enough bunch. They don't talk much though, but always complaining about our Nuclear weapons, claiming they affect parallel universes every time they're triggered".

The disclosure comes after 70 years of countless sightings and abductions, raising questions as to why it is they are here.

"The aliens are actually harmless and only interested in the planet's natural resources," the briefing concluded, "which shouldn't cause us any problems whatsoever".

In their disclaimer notice, Waterford Whispers, states that the site is a about "fabricated satirical newspaper":

Waterford Whispers News is a fabricated satirical newspaper and comedy website published by Waterford Whispers News.

Waterford Whispers News uses invented names in all its stories, except in cases when public figures are being satirized. Any other use of real names is accidental and coincidental.

Waterford Whispers is largely recognized as an Irish counterpart to *The Onion* among its primary reader base (in the UK and Ireland). However, previous items from the site have been confused for real news, including reports that the Pope commissioned J.K. Rowling to rewrite the Bible, the Muppet known as "Animal" had died, and that the Vatican decreed Jesus was not returning. *Disclose.tv* has passed on a decent share of fake news items, including claims a baby in the Philippines was born with Stigmata and Edward Snowden had been "reported dead by his girlfriend."

NASA About to Confirm They Have Discovered Alien Life? Or Just More Lies and Waffle?

NASA is apparently primed to reveal its latest set of major discoveries on Monday after scanning planets outside our solar system for life. But many do believe that this could also be yet just more 'drip feeding' of disclosure and no real confirmation of anything – Are NASA playing games with us and the TRUTH?

Many people worldwide are now on the edge of their seats amid hopes the new find could shed light on the search for aliens. Although we are used to disappointment from NASA and being given information which does not really confirm anything.

NASA has been questing to find the first traces of life to prove man is not alone in the universe. Just recently we reported on NASA's latest find of a solar system like ours, where there could be the potential for life – please check out our article below:

The discoveries will come from the team at the Kepler space telescope which looks for habitable worlds outside the Solar System.

The telescope searches for Earth-sized planets in the "habitable zones" of nearby stars – and has found thousands of planets which could harbor life. Speculation is mounting about the announcement by NASA – which is set for **3pm on Monday, June 19.**

Infested with life out there – And does NASA know about it?

As mentioned in our previous article above, last year, NASA announced the incredible find of nine planets orbiting the star TRAPPIST-1. Three of the planets were believed to be in the star's habitable zone and capable of harboring life.

Scientists are launching a new satellite – the James Webb Telescopes – to study seven of the new worlds! But again, NASA has not mentioned anything about Alien Beings potentially living on any of these planets and gave excuses as to why that may not be possible (as always!).

A spokesman for NASA said:

"The latest Kepler catalog of planet candidates was created using the most sophisticated analyses yet, yielding the most complete and reliable accounting of distant worlds to date.

"This survey will enable new lines of research in exoplanet study, which looks at planets outside our solar system."

U.I.P SUMMARY

This is not the first and not the last NASA announcement about their search for habitable planets, but one thing is for sure, these announcements appear to be happening more frequently – The drip feeding of disclosure is intensifying it appears?!

NASA know very well that we are NOT alone in the universe and are also aware that we are being visited by numerous different ET races. Why do they not just tell us this then if that's that case, I hear you shout?! Well, it appears that the Elite fear people like you and I knowing that there are advanced beings visiting us, as they know the kind of opportunities that will arise once humans has accepted their existence – the Elites false borders and war mongering will be obliterated very quickly!

The chances of NASA announcing on Monday is incredibly slim, but please ensure that you listen to their every word, as some where in-between the lines are the TRUTH about what they really know!!!

Please check out below a couple of our recent articles on why the Elite do not want us to know the truth about the ET's and an article on 'Tabby's star' which is thought to have a huge Alien megastructure around it.

Interesting times people.

UFO International Project » Analysis & Implications » Scientists Confirm That Newly discovered STAR system, could be home to interplanetary ALIENS (Drip Feeding Disclosure)

« Planet Earth, An Alien Prison? Leading Scientists Now Believe So!

Man Bundled into Van at Area 51, After Stealing Proof of ALIENS »

Scientists Confirm That Newly discovered STAR system, could be home to interplanetary ALIENS (Drip Feeding Disclosure)

It appears that we are now being fed some huge amounts of drip feeding of Disclosure, after the newly discovered TRAPPIST-1 star and its nearby planets could be home to interplanetary alien life, leading scientists have now claimed – not long after NASA announced this exciting find deep in space.

Back in February 2017, Nasa announced the discovery of a star known as TRAPPIST-1 which was found to have seven planets orbiting it. This newly discovered solar system, which is the closest to have been found to Earth, piqued the interest of alien hunters, as three of its planets fall into the star's habitable zone – the region around a host star where conditions are neither too hot nor too cold to support life – in fact many believe that these planets could have even more advanced beings on them due to such impressive surroundings!?

A new study from Harvard University suggests that alien life could travel between the planets due to their proximity to one another, suggesting that the planets alignment could have created far more advanced Beings living on them – just imagine being able to travel to a habitable planet within your very own solar system??

What lies beyond?

The researchers say that when an asteroid or comet or any form of space rock collides into one of the planets, it may propel dirt into space which carries microbial life – spreading life throughout the Universe!

Since the planets are near to each other, the dirt could land on the next planet, essentially transporting life between them! This is an interesting theory known as panspermia and many scientists believe that is potentially how life on Earth began – when an asteroid collided with Mars, causing the supposedly once microbial-full soil to be flung Earthbound – many also believe that Mars was once a habitable planet with ET Beings on it, that is until come kind of 'disaster' struck!?

Was Life on Mars Destroyed by Some Catastrophic Event!?

Harvard astronomer Avi Loeb told Gizmodo said:

"The rocks are driven into space.

"If there is life on one of them, life may be preserved inside these rocks and be transferred to another planet."

Prof Loeb went on to say that the likelihood of this happening is high, based on models of how life has been transported between islands on Earth – it appears that the Elite are trying to drop us some huge hints??

He then added:

"These planets are like islands on the surface of the Earth, and there are studies of the immigration of species from one island to another.

"We used the same model to illustrate that the likelihood is very

high for transfer of life."

This is all perfectly timed after other leading scientists confirmed that recent signal blasts from space could be caused by advance Extraterrestrial Beings – please click here for more on this:

Official Alien Contact and Humanity's Path to The Universe

There is a clear defined reason why the Elite of this world do not want us to have Extra-terrestrial Disclosure and it is the same reason why our so-called leaders will not tell us the truth – they do NOT want to lose the Control over us!

If the friendly Aliens want to play nicely with us, disclosure of their existence will bring with it a possible doorway to the Universe, an opportunity for us to explore the darkness between the stars and beyond. No longer will planet earths borders between countries appear necessary or relevant, and our world leaders and religious leaders would not appear so significant in our lives, in many aspects they may even come across as 'fakes' and 'un-important'. And THIS is what the elite who manipulate this planet are afraid of.

The Elites end of days?

UFO sightings are dramatically on the rise and have been for a while (please click HERE for more). The ET's want and need to be seen and many people believe that the Alien Beings are trying to force their own disclosure. Hand in hand with the increased 'clear' UFO sightings, is a world-scaled awakening throughout humans, it is no longer a case of "are we alone in the Universe" but instead the question is "when will we discover who the Aliens are and what do they want?".

The Elite are more than aware of this Global awakening so have put into place plans with other agencies/corporations' ways of 'delaying' disclosure and in some cases 'drip feeding us with disclosure' perhaps for when the Aliens force their own disclosure, so our world leaders etc. can turn around and tell us "that we did try telling you!".

We are quite often TEASED by NASA about the existence of life in the Universe, by means of news headlines about 'Life on Mars' (please click HERE) and in some cases even delayed acceptance statements of there being other advanced life forms out there in the Universe. We have recently been informed by NASA that they will discover Alien life forms in the next ten years, which appears to be a fantastic delay tactic in itself (please click HERE for more info on this). "We are on the cusp of discovering alien civilizations", NASA's top scientists have said. They have very recently predicted that we are just one generation away from finding something in our Milky Way neighborhood, which is bustling with environments conducive to life……. let the Drip Feeding of Disclosure begin!

'We will discover Alien life by 2025' – Dr. Ellen Stofan from NASA helping to 'drip feed the truth'

Even the UN have been hinting at what's to come in a recent strange advertisement of theirs – please see video below:

The list could go on and on of the ways that the elite are drip feeding us all with disclosure, from adverts to Astronauts and even the US President getting heavily involved in this incredibly important subject matter, but all of this in the greater scheme of things is irrelevant – The Aliens are already HERE and have been for a long time and those who used to control us are soon to become surplus to requirement, BIG exciting changes are coming for us all!

What Will the Aliens Bring?

No matter which way you look at it this planet is in turmoil, what with conflicts and tensions between countries going on all around the world, created by war-mongering power-hungry leaders! Planet earth's natural resources are being drastically drained daily for money and 'control', a divide amongst different religions and races is being ridiculously encouraged by the Elite (again to CONTROL us) and moreover the people of this world appear to brainwash into submission that money, TV and who owns the best gadgets is what is important in life.

BUT, there are big changes ahead, something which many of us have opened our minds up too, and these changes are going to inevitably happen across the planet when the Extra-terrestrials officially announce their presence to us all.

The Aliens aren't coming, they are already HERE!

At first you will have a clear divide amongst the masses on planet earth, of those who welcome our ET friends, and those who are threatened by them. Of course, the world powers will try to get the people on side to fear these unknown 'demons' from the sky and with this will come violence from certain fractions of humanity. Some religious groups will oppose the ET's and try to discourage the world from uniting with them. Those who have a false sense of security and have been absorbed by the Elites brainwashing will potentially go up in arms and be untrusting of our ET visitors.

The ET's (if it's the good ones that appear) will probably first announce themselves to the world of who they are and where they are from, and their reasons for their official contact with Planet Earth. As soon as most of humans accept who are out ET visitors are and why they are here, a sudden worldwide 'full' awakening will start to begin, and the universe will suddenly be deemed as a new playground of discovery for the people of this world.

Our technology will become obsolete in comparison to what the Aliens use and have further knowledge of. It has been discussed by many leading researchers that in the greater scheme of things in the Universe humans is a mere type ZERO civilization, since we are still very dependable on fossil fuels and have not yet quite understood the full benefits of using our local star as a true energy resource. Compared to MANY other ET Beings out there who have been around in the Universe a lot longer than what we have, we would be pretty much prehistoric-like in comparison to them!

We Are a Mere Type 0 Civilization – Against Types I, II, III, IV And V, ALIEN Civilizations believes many leading researchers around the world. If there are planets out there with life on it that has been around the Universe for millions of years over what we have, it is obvious that their technology and Science knowledge would be FAR superior to our own. The 'Kardashev scale' has been developed as a way of measuring a civilization's technological advancement based upon how much usable energy it has at its disposal – hence we are classed as a Type of

Civilization. Many leading researchers believe that as the populations grows and expands outwards its energy requirements will increase dramatically, what with the requirements of its various technological machines. Therefore, the Kardashev Scale was created, a way of measuring a civilization's technological advancement based upon how much usable energy it has at its disposal.

Bibliography

Baumrind, Jed & Krulwich, Robert. "Carl Sagan and Ann Druyan's Ultimate Mix Tape." National Public Radio. Feb. 12, 2010. (Feb. 1, 2012) http://www.npr.org/2010/02/12/123534818/carl-sagan-and-ann-druyans-ultimate-mix-tape

Anderson, Charlene. "Projects: Observing Earth. Carl Sagan on Venus and Mars." (Feb. 1, 2012) http://planetary.org/programs/projects/observing_earth/sagan_climate_change.html

Brazell, Karen. "Happy Belated Birthday, Carl!" PBS. Nov. 10, 2010. (Feb. 7, 2012) http://www.pbs.org/wnet/need-to-know/the-daily-need/photo-happy-belated-birthday-carl/5026/

Davidson, Keay. "Carl Sagan: A Life." John Wiley & Sons, Inc. 1999.

Druyan, Ann. "Carl Sagan." Personal interview. Feb. 2, 2012.

International Space Hall of Fame. "Carl Sagan." (Feb. 5, 2012) http://www.nmspacemuseum.org/halloffame/detail.php?id=149

Morrison, David. "Carl Sagan's Life and Legacy as a Scientist, Teacher and Skeptic." Skeptical Inquirer. Vol. 31.1, January 2007. (Feb. 1, 2012) http://www.csicop.org/si/show/carl_sagans_life_and_legacy_as_scientist_teacher_and_skeptic

NASA. "Carl Sagan: Solar System Exploration." April 12, 2011. (Feb. 5, 2012) http://solarsystem.nasa.gov/people/profile.cfm?Code=SaganC

NASA Jet Propulsion Laboratory. "What is the Golden Record?" (Feb. 1, 2012) http://voyager.jpl.nasa.gov/spacecraft/goldenrec.html

National Archives. "Unidentified Flying Objects: Project BLUE BOOK." Freedom of Information Act. (Feb. 5, 2012) http://www.archives.gov/foia/ufos.html

The Planetary Society. "Carl Sagan: Founder and First President." (Feb. 5, 2012) http://planetary.org/about/founders/carl_sagan.html

The Planetary Society."Visions of Mars: A Message to the Future." (Feb. 1, 2012). http://www.planetary.org/programs/projects/messages/vom.html

The Planetary Society. "Visions of Mars Greetings: Carl Sagan." (Feb. 1, 2012). http://www.planetary.org/special/vomgreetings/greet2/SAGAN.html

Poundstone, William. "Carl Sagan: A Life in the Cosmos." Henry Holt and Co. 1999.

Poundstone, William. "Carl Sagan." Personal Interview. Jan. 31, 2012.

Quarles, Norma. "Carl Sagan Dies at 62." CNN. Dec. 20, 1996. (Feb. 7, 2012) http://articles.cnn.com/1996-12-20/us/9612_20_sagan_1_bone-marrow-blood-disease-planets?_s=PM:US

Sagan, Carl and George Mullen. "Earth and Mars: Evolution of Atmospheres and Surface Temperatures." Science. Vol. 177, no. 4043. 1972. (Feb. 1, 2012) http://www.sciencemag.org/content/177/4043/52

SETI. "Carl Sagan Center." (Feb. 1, 2012) http://www.seti.org/carlsagancenter

Turco, R.P., Toon, O.B., Ackerman, T.P., Pollack, J.B., & Sagan,

Carl. "Nuclear Winter: Global Consequences of Multiple Nuclear Explosions." Science. Vol. 222, no. 4630. 1983. (Feb. 1, 2012) http://www.sciencemag.org/content/222/4630/1283.shortU.S. Environmental Protection Agency. "Carbon dioxide." April 15, 2011. (Feb. 1, 2012) http://www.epa.gov/climatechange/emissions/co2.html

Davies, Paul (18 November 2013). "Are We Alone in the Universe?". The New York Times. Retrieved 20 November 2013.

Filkin, David; Hawking, Stephen W. (1998). Stephen Hawking's universe: the cosmos explained. Art of Mentoring Series. Basic Books. p. 194. ISBN 0-465-08198-3.

Rauchfuss, Horst (2008). Chemical Evolution and the Origin of Life. trans. Terence N. Mitchell. Springer. ISBN 3-540-78822-0.

Loeb, Abraham (October 2014). "The Habitable Epoch of the Early Universe". International Journal of Astrobiology. 13 (4): 337–339. arXiv:1312.0613 Freely accessible. Bibcode:2014IJAsB..13..337L. doi:10.1017/S1473550414000196. Retrieved 15 December 2014.

Dreifus, Claudia (2 December 2014). "Much-Discussed Views That Go Way Back – Avi Loeb Ponders the Early Universe, Nature and Life". The New York Times. Retrieved 3 December 2014.

Rampelotto, P. H. (April 2010). Panspermia: A Promising Field of Research (PDF). Astrobiology Science Conference 2010: Evolution and Life: Surviving Catastrophes and Extremes on Earth and Beyond. 20–26 April 2010. League City, Texas. Bibcode:2010LPICo1538.5224R.

Gonzalez, Guillermo; Richards, Jay Wesley (2004). The privileged planet: how our place in the cosmos is designed for discovery. Regnery Publishing. pp. 343–345. ISBN 0-89526-065-4.

Moskowitz, Clara (29 March 2012). "Life's Building Blocks May Have Formed in Dust Around Young Sun". Space.com. Retrieved 30 March 2012.

Choi, Charles Q. (21 March 2011). "New Estimate for Alien Earths: 2 Billion in Our Galaxy Alone". Space.com. Retrieved 24 April 2011.

Torres, Abel Mendez (26 April 2013). "Ten potentially habitable exoplanets now". Habitable Exoplanets Catalog. University of Puerto Rico. Retrieved 29 April 2013.

Overbye, Dennis (4 November 2013). "Far-Off Planets Like the Earth Dot the Galaxy". The New York Times. Retrieved 5 November 2013.

Petigura, Eric A.; Howard, Andrew W.; Marcy, Geoffrey W. (31 October 2013). "Prevalence of Earth-size planets orbiting Sun-like stars". Proceedings of the National Academy of Sciences of the United States of America. 110 (48): 19273–19278. arXiv:1311.6806 Freely accessible. Bibcode:2013PNAS..11019273P. doi:10.1073/pnas.1319909110. PMC 3845182 Freely accessible. PMID 24191033. Retrieved 5 November 2013.

Khan, Amina (4 November 2013). "Milky Way may host billions of Earth-size planets". Los Angeles Times. Retrieved 5 November 2013.

Hoehler, Tori M.; Amend, Jan P.; Shock, Everett L. (2007). "A "Follow the Energy" Approach for Astrobiology". Astrobiology. 7 (6): 819–823. Bibcode:2007AsBio...7..819H. doi:10.1089/ast.2007.0207. ISSN 1531-1074. PMID 18069913.

Jones, Eriita G.; Lineweaver, Charles H. (2010). "To What Extent Does Terrestrial Life "Follow the Water"?" (PDF). Astrobiology.

10 (3): 349–361. Bibcode:2010AsBio..10..349J. doi:10.1089/ast.2009.0428. ISSN 1531-1074.

"Aliens may be more like us than we think". University of Oxford. 31 October 2017.

Stevenson, David S.; Large, Sean (25 October 2017). "Evolutionary exobiology: Towards the qualitative assessment of biological potential on exoplanets". International Journal of Astrobiology: 1–5. doi:10.1017/S1473550417000349.

Bond, Jade C.; O'Brien, David P.; Lauretta, Dante S. (June 2010). "The Compositional Diversity of Extrasolar Terrestrial Planets. I. In Situ Simulations". The Astrophysical Journal. 715 (2): 1050–1070. arXiv:1004.0971 Freely accessible. Bibcode:2010ApJ...715.1050B. doi:10.1088/0004-637X/715/2/1050.

Pace, Norman R. (20 January 2001). "The universal nature of biochemistry". Proceedings of the National Academy of Sciences of the United States of America. 98 (3): 805–808. Bibcode:2001PNAS...98..805P. doi:10.1073/pnas.98.3.805. PMC 33372 Freely accessible. PMID 11158550.

National Research Council (2007). "6.2.2: Nonpolar Solvents". The Limits of Organic Life in Planetary Systems. The National Academies Press. p. 74. doi:10.17226/11919. ISBN 978-0-309-10484-5.

Nielsen, Forrest H. (1999). "Ultrarace Minerals". In Shils, Maurice E.; Shike, Moshe. Modern Nutrition in Health and Disease (9th ed.). Williams & Wilkins. pp. 283–303. ISBN 978-0-683-30769-6.

Mix, Lucas John (2009). Life in space: astrobiology for everyone. Harvard University Press. p. 76. ISBN 0-674-03321-3. Retrieved 8 August 2011.

Horowitz, Norman H. (1986). To Utopia and Back: The Search for

Life in the Solar System. W. H. Freeman & Co. ISBN 0-7167-1765-4.

Dyches, Preston; Chou, Felcia (7 April 2015). "The Solar System and Beyond is Awash in Water". NASA. Retrieved 8 April 2015.

Hays, Lindsay, ed. (2015). "NASA Astrobiology Strategy 2015" (PDF). NASA. p. 65.

Summons, Roger E.; Amend, Jan P.; Bish, David; Buick, Roger; Cody, George D.; Des Marais, David J.; Dromart, Gilles; Eigenbrode, Jennifer L.; et al. (2011). "Preservation of Martian Organic and Environmental Records: Final Report of the Mars Biosignature Working Group". Astrobiology. 11 (2): 157–81. Bibcode:2011AsBio..11..157S. doi:10.1089/ast.2010.0506. PMID 21417945. "There is general consensus that extant microbial life on Mars would probably exist (if at all) in the subsurface and at low abundance."

Michalski, Joseph R.; Cuadros, Javier; Niles, Paul B.; Parnell, John; Deanne Rogers, A.; Wright, Shawn P. (2013). "Groundwater activity on Mars and implications for a deep biosphere". Nature Geoscience. 6 (2): 133–8. Bibcode:2013NatGe...6..133M. doi:10.1038/ngeo1706.

"Habitability and Biology: What are the Properties of Life?". Phoenix Mars Mission. The University of Arizona. Retrieved 6 June 2013. "If any life exists on Mars today, scientists believe it is most likely to be in pockets of liquid water beneath the Martian surface."

Tritt, Charles S. (2002). "Possibility of Life on Europa". Milwaukee School of Engineering. Archived from the original on 9 June 2007. Retrieved 10 August 2007.

Kargel, Jeffrey S.; Kaye, Jonathan Z.; Head, James W.; Marion, Giles M.; Sassen, Roger; et al. (November 2000). "Europa's Crust and Ocean: Origin, Composition, and the Prospects for Life". Icarus. 148 (1): 226–265. Bibcode:2000Icar..148..226K. doi:10.1006/icar.2000.6471.

Schulze-Makuch, Dirk; Irwin, Louis N. (2001). "Alternative Energy Sources Could Support Life on Europa" (PDF). Departments of Geological and Biological Sciences, University of Texas at El Paso. Archived from the original (PDF) on 3 July 2006. Retrieved 21 December 2007.

O'Leary, Margaret R. (2008). Anaxagoras and the Origin of Panspermia Theory. iUniverse. ISBN 978-0-595-49596-2.

Berzelius, Jöns Jacob (1834). "Analysis of the Alais meteorite and implications about life in other worlds". Annalen der Chemie und Pharmacie. 10: 134–135.

Thomson, William (August 1871). "The British Association Meeting at Edinburgh". Nature. 4 (92): 261–278. Bibcode:1871Natur...4..261.. doi:10.1038/004261a0. PMC 2070380 Freely accessible. "We must regard it as probably to the highest degree that there are countless seed-bearing meteoritic stones moving through space."

Demets, René (October 2012). "Darwin's Contribution to the Development of the Panspermia Theory". Astrobiology. 12 (10): 946–950. Bibcode:2012AsBio..12..946D. doi:10.1089/ast.2011.0790. PMID 23078643.

Arrhenius, Svante (March 1908). Worlds in the Making: The Evolution of the Universe. trans. H. Borns. Harper & Brothers. OCLC 1935295.

Hoyle, Fred; Wickramasinghe, Chandra; Watson, John (1986). Viruses from Space and Related Matters (PDF). University College Cardiff Press. Bibcode: 1986vfsr.book.....H. ISBN 978-0-906449-93-6.

Crick, F. H.; Orgel, L. E. (1973). "Directed Panspermia". Icarus. 19 (3): 341–348. Bibcode:1973Icar...19..341C. doi:10.1016/0019-1035(73)90110-3.

Orgel, L. E.; Crick, F. H. (January 1993). "Anticipating an RNA world. Some past speculations on the origin of life: Where are they today?". FASEB Journal. 7 (1): 238–239. PMID 7678564.

Clark, Stuart (26 September 2003). "Acidic clouds of Venus could harbor life". New Scientist. Retrieved 30 December 2015.

Redfern, Martin (25 May 2004). "Venus clouds 'might harbor life'". BBC News. Retrieved 30 December 2015.

Dartnell, Lewis R.; Nordheim, Tom Andre; Patel, Manish R.; Mason, Jonathon P.; et al. (September 2015). "Constraints on a potential aerial biosphere on Venus: I. Cosmic rays". Icarus. 257: 396–405. Bibcode:2015Icar. 257..396D. doe: 10.1016/j.icarus.2015.05.006. Retrieved 20 August 2015.

"Did the Early Venus Harbor Life? (Weekend Feature)". The Daily Galaxy. 2 June 2012. Retrieved 22 May 2016.

"Was Venus once a habitable planet?". European Space Agency. 24 June 2010. Retrieved 22 May 2016.

Atkinson, Nancy (24 June 2010). "Was Venus once a water world?". Universe Today. Retrieved 22 May 2016.

Bortman, Henry (26 August 2004). "Was Venus Alive? 'The Signs are Probably There'". Space.com. Retrieved 22 May 2016.

Ojha, L.; Wilhelm, M. B.; Murchie, S. L.; McEwen, A. S.; Wray, J. J.; Hanley, J.; Massé, M.; Chojnacki, M. (2015). "Spectral evidence for hydrated salts in recurring slope lineae on Mars". Nature Geoscience. 8 (11): 829–832. Bibcode:2015NatGe...8..829O. doi:10.1038/ngeo2546.

"Top 10 Places to Find Alien Life: Discovery News".

News.discovery.com. 8 June 2010. Retrieved 13 June 2012.

Baldwin, Emily (26 April 2012). "Lichen survives harsh Mars environment". Skamania News. Retrieved 27 April 2012.

Kohler, Ulrich (26 April 2012). "The adaptation potential of extremophiles to Martian surface conditions and its implication for the habitability of Mars" (PDF). European Geosciences Union. Archived from the original (PDF) on 8 June 2012. Retrieved 27 April 2012.

Chang, Kenneth (9 December 2013). "On Mars, an Ancient Lake and Perhaps Life". The New York Times. Retrieved 9 December 2013.

"Science – Special Collection – Curiosity Rover on Mars". Science. 9 December 2013. Retrieved 9 December 2013.

Grotzinger, John P. (24 January 2014). "Introduction to Special Issue – Habitability, Taphonomy, and the Search for Organic Carbon on Mars". Science. 343 (6169): 386–387. Bibcode:2014Sci... 343..386G. doi:10.1126/science.1249944. PMID 24458635. Retrieved 24 January 2014.

"Special Issue – Table of Contents – Exploring Martian Habitability". Science. 343 (6169): 345–452. 24 January 2014. Retrieved 24 January 2014.

"Special Collection – Curiosity – Exploring Martian Habitability". Science. 24 January 2014. Retrieved 24 January 2014.

Grotzinger, J. P.; et al. (24 January 2014). "A Habitable Fluvio-Lacustrine Environment at Yellowknife Bay, Gale Crater, Mars". Science. 343 (6169): 1242777. Bibcode:2014Sci...343A.386G. doi:10.1126/science.1242777. PMID 24324272. Retrieved 24 January 2014.

Küppers, M.; O'Rourke, L.; Bockelée-Morvan, D.; Zakharov, V.;

Lee, S.; Von Allmen, P.; Carry, B.; Teyssier, D.; Marston, A.; Müller, T.; Crovisier, J.; Barucci, M. A.; Moreno, R. (23 January 2014). "Localized sources of water vapour on the dwarf planet (1) Ceres". Nature. 505 (7484): 525–527. Bibcode:2014Natur. 505..525K. doi:10.1038/nature12918. ISSN 0028-0836. PMID 24451541.

Campins, H.; Comfort, C. M. (23 January 2014). "Solar system: Evaporating asteroid". Nature. 505 (7484): 487–488. Bibcode:2014Natur. 505..487C. doi:10.1038/505487a. PMID 24451536.

A'Hearn, Michael F.; Feldman, Paul D. (1992). "Water vaporization on Ceres". Icarus. 98 (1): 54–60. Bibcode:1992Icar...98...54A. doi:10.1016/0019-1035(92)90206-M.

Duffy, Alan (15 June 2015). "What on Ceres are those bright spots?". Cosmos.

Rivkin, Andrew (21 July 2015). "Dawn at Ceres: A haze in Occator crater?". The Planetary Society. Retrieved 24 July 2015.

O'Neill, Ian (5 March 2009). "Life on Ceres: Could the Dwarf Planet be the Root of Panspermia". Universe Today. Retrieved 30 January 2012.

Catling, David C. (2013). Astrobiology: A Very Short Introduction. Oxford: Oxford University Press. p. 99. ISBN 0-19-958645-4.

Boyle, Alan (22 January 2014). "Is There Life on Ceres? Dwarf Planet Spews Water Vapor". NBC. Retrieved 10 February 2015.

Ponnamperuma, Cyril; Molton, Peter (January 1973). "The prospect of life on Jupiter". Space Life Sciences. 4 (1): 32–44. Bibcode:1973SLSci...4...32P. doi:10.1007/BF02626340. PMID 4197410.

Irwin, Louis Neal; Schulze-Makuch, Dirk (June 2001). "Assessing the Plausibility of Life on Other Worlds". Astrobiology. 1 (2): 143–160. Bibcode:2001AsBio...1..143I. doi:10.1089/153110701753198918. PMID 12467118.

Dyches, Preston; Brown, Dwayne (12 May 2015). "NASA Research Reveals Europa's Mystery Dark Material Could Be Sea Salt". NASA. Retrieved 12 May 2015.

"NASA's Hubble Observations Suggest Underground Ocean on Jupiter's Largest Moon". NASA News. 12 March 2015. Retrieved 15 March 2015.

"Jupiter moon Ganymede could have ocean with more water than Earth – NASA". Russia Today (RT). 13 March 2015. Retrieved 13 March 2015.

Clavin, Whitney (1 May 2014). "Ganymede May Harbor 'Club Sandwich' of Oceans and Ice". NASA. Jet Propulsion Laboratory. Retrieved 1 May 2014.

Vance, Steve; Bouffard, Mathieu; Choukroun, Mathieu; Sotina, Christophe (12 April 2014). "Ganymede's internal structure including thermodynamics of magnesium sulfate oceans in contact with ice". Planetary and Space Science. 96: 62–70. Bibcode:2014P&SS...96...62V. doi: 10.1016/j.pss.2014.03.011. Retrieved 2 May 2014.

"Video (00:51) – Jupiter's 'Club Sandwich' Moon". NASA. 1 May 2014. Retrieved 2 May 2014.

Chang, Kenneth (12 March 2015). "Suddenly, It Seems, Water Is Everywhere in Solar System". The New York Times. Retrieved 12 March 2015.

Kuskov, O. L.; Kronrod, V. A. (2005). "Internal structure of Europa and Callisto". Icarus. 177 (2): 550–569. Bibcode:2005Icar..177..550K. doi: 10.1016/j.icarus.2005.04.014.

Showman, Adam P.; Malhotra, Renu (1999). "The Galilean Satellites" (PDF). Science. 286 (5437): 77–84. doi:10.1126/science.286.5437.77. PMID 10506564.

Hsiao, Eric (2004). "Possibility of Life on Europa" (PDF). University of Victoria.

Friedman, Louis (14 December 2005). "Projects: Europa Mission Campaign". The Planetary Society. Archived from the original on 11 August 2011. Retrieved 8 August 2011.

Atkinson, Nancy (2009). "Europa Capable of Supporting Life, Scientist Says". Universe Today. Retrieved 18 August 2011.

Plait, Phil (17 November 2011). "Huge lakes of water may exist under Europa's ice". Discover. Bad Astronomy Blog.

"Scientists Find Evidence for "Great Lake" on Europa and Potential New Habitat for Life". The University of Texas at Austin. 16 November 2011.

Cook, Jia-Rui C. (11 December 2013). "Clay-Like Minerals Found on Icy Crust of Europa". NASA. Retrieved 11 December 2013.

Wall, Mike (5 March 2014). "NASA hopes to launch ambitious mission to icy Jupiter moon". Space.com. Retrieved 15 April 2014.

Clark, Stephen (14 March 2014). "Economics, water plumes to drive Europa mission study". Spaceflight Now. Retrieved 15 April 2014.

Coustenis, A.; et al. (March 2009). "TandEM: Titan and Enceladus mission". Experimental Astronomy. 23 (3): 893–946. Bibcode:2009ExA....23..893C. doi:10.1007/s10686-008-9103-z.

Lovett, Richard A. (31 May 2011). "Enceladus named sweetest spot for alien life". Nature. Nature. doi:10.1038/news.2011.337. Retrieved 3 June 2011.

Then, Ker (13 September 2005). "Scientists Reconsider Habitability of Saturn's Moon". Space.com.

Britt, Robert Roy (28 July 2006). "Lakes Found on Saturn's Moon Titan". Space.com.

"Lakes on Titan, Full-Res: PIA08630". 24 July 2006. Archived from the original on 29 September 2006.

"What is Consuming Hydrogen and Acetylene on Titan?". NASA/JPL. 2010. Archived from the original on 29 June 2011. Retrieved 6 June 2010.

Strobel, Darrell F. (2010). "Molecular hydrogen in Titan's atmosphere: Implications of the measured tropospheric and thermosphere mole fractions". Icarus. 208 (2): 878–886. Bibcode:2010Icar..208..878S. doi: 10.1016/j.icarus.2010.03.003.

McKay, C. P.; Smith, H. D. (2005). "Possibilities for methanogenic life in liquid methane on the surface of Titan". Icarus. 178 (1): 274–276. Bibcode:2005Icar..178..274M. doi: 10.1016/j.icarus.2005.05.018.

Hoyle, Fred (1982). Evolution from Space (The Omni Lecture) and Other Papers on the Origin of Life. Enslow. pp. 27–28. ISBN 0-89490-083-8.

Hoyle, Fred; Wickramasinghe, Chandra (1984). Evolution from Space: A Theory of Cosmic Creationism. Simon & Schuster. ISBN 0-671-49263-2.

Hoyle, Fred (1985). Living Comets. Cardiff: University College, Cardiff Press.

Wickramasinghe, Chandra (June 2011). "Viva Panspermia". The Observatory.

Wesson, P (2010). "Panspermia, Past and Present: Astrophysical and Biophysical Conditions for the Dissemination of Life in

Space". Sp. Sci.Rev. 1–4. 156: 239–252. arXiv:1011.0101 Freely accessible. Bibcode:2010SSRv..156..239W. doi:10.1007/s11214-010-9671-x.

Hussmann, Hauke; Sohl, Frank; Spohn, Tilman (November 2006). "Subsurface oceans and deep interiors of medium-sized outer planet satellites and large trans-neptunian objects" (PDF). Icarus. 185 (1): 258–273. Bibcode:2006Icar..185..258H. doi:10.1016/j.icarus.2006.06.005.

Choi, Charles Q. "The Chance for Life on Io". Retrieved 25 May 2013.

Crenson, Matt (6 August 2006). "Experts: Little Evidence of Life on Mars". Associated Press. Archived from the original on 16 April 2011. Retrieved 8 March 2011.

McKay, David S.; Gibson, Everett K., Jr.; Thomas-Keprta, Kathie L.; Vali, Hojatollah; Romanek, Christopher S.; et al. (August 1996). "Search for Past Life on Mars: Possible Relic Biogenic Activity in Martian Meteorite ALH84001". Science. 273 (5277): 924–930. Bibcode:1996Sci...273..924M. doi:10.1126/science.273.5277.924. PMID 8688069.

McKay, David S.; Thomas-Keprta, Kathy L.; Clemett, Simon J.; Gibson, Everett K., Jr.; Spencer, Lauren; Wentworth, Susan J. (August 2009). "Life on Mars: New Evidence from Martian Meteorites". Proceedings of the SPIE. Instruments and Methods for Astrobiology and Planetary Missions XII. 7441. 744102. Bibcode:2009SPIE.7441E..02M. doi:10.1117/12.832317.

Webster, Guy (27 February 2014). "NASA Scientists Find Evidence of Water in Meteorite, Reviving Debate Over Life on Mars". NASA. Retrieved 27 February 2014.

White, Lauren M.; Gibson, Everett K.; Thomnas-Keprta, Kathie L.; Clemett, Simon J.; McKay, David (19 February 2014).

"Putative Indigenous Carbon-Bearing Alteration Features in Martian Meteorite Yamato 000593". Astrobiology. 14 (2): 170–181. Bibcode:2014AsBio..14..170W. doi:10.1089/ast.2011.0733. PMC 3929347 Freely accessible. PMID 24552234. Retrieved 27 February 2014.

Gannon, Megan (28 February 2014). "Mars Meteorite with Odd 'Tunnels' & 'Spheres' Revives Debate Over Ancient Martian Life". Space.com. Retrieved 28 February 2014.

Chambers, Paul (1999). Life on Mars; The Complete Story. London: Blandford. ISBN 0-7137-2747-0.

Klein, Harold P.; Levin, Gilbert V.; Levin, Gilbert V.; Oyama, Vance I.; Lederberg, Joshua; Rich, Alexander; Hubbard, Jerry S.; Hobby, George L.; Straat, Patricia A.; Berdahl, Bonnie J.; Carle, Glenn C.; Brown, Frederick S.; Johnson, Richard D. (1 October 1976). "The Viking Biological Investigation: Preliminary Results". Science. 194 (4260): 99–105. Bibcode:1976Sci...194...99K. doi:10.1126/science.194.4260.99. PMID 17793090. Retrieved 15 August 2008.

Beegle, Luther W.; Wilson, Michael G.; Abilleira, Fernando; Jordan, James F.; Wilson, Gregory R. (August 2007). "A Concept for NASA's Mars 2016 Astrobiology Field Laboratory". Astrobiology. 7 (4): 545–577. Bibcode:2007AsBio...7..545B. doi:10.1089/ast.2007.0153. PMID 17723090. Retrieved 20 July 2009.

"ExoMars rover". ESA. Retrieved 14 April 2014.

Berger, Brian (2005). "Exclusive: NASA Researchers Claim Evidence of Present Life on Mars".

"NASA denies Mars life reports". spacetoday.net. 2005.

Spotts, Peter N. (28 February 2005). "Sea boosts hope of finding signs of life on Mars". The Christian Science Monitor. Retrieved

18 December 2006.

Chow, Dennis (22 July 2011). "NASA's Next Mars Rover to Land at Huge Gale Crater". Space.com. Retrieved 22 July 2011.

Amos, Jonathan (22 July 2011). "Mars rover aims for deep crater". BBC News. Retrieved 22 July 2011.

Glaser, Linda (27 January 2015). "Introducing: The Carl Sagan Institute". Archived from the original on 27 February 2015. Retrieved 11 May 2015.

"Carl Sagan Institute – Research". May 2015. Retrieved 11 May 2015.

Cofield, Calla (30 March 2015). "Catalog of Earth Microbes Could Help Find Alien Life". Space.com. Retrieved 11 May 2015.

Callahan, M.P.; Smith, K.E.; Cleaves, H.J.; Ruzica, J.; Stern, J.C.; Glavin, D.P.; House, C.H.; Dworkin, J.P. (11 August 2011). "Carbonaceous meteorites contain a wide range of extraterrestrial nucleobases". Proceedings of the National Academy of Sciences. PNAS. 108 (34): 13995–13998. Bibcode:2011PNAS..10813995C. doi:10.1073/pnas.1106493108. PMC 3161613 Freely accessible. PMID 21836052. Retrieved 15 August 2011.

Steigerwald, John (8 August 2011). "NASA Researchers: DNA Building Blocks Can Be Made in Space". NASA. Retrieved 10 August 2011.

"DNA Building Blocks Can Be Made in Space, NASA Evidence Suggests". ScienceDaily. 9 August 2011. Retrieved 9 August 2011.

Chow, Denise (26 October 2011). "Discovery: Cosmic Dust Contains Organic Matter from Stars". Space.com. Retrieved 26 October 2011.

"Astronomers Discover Complex Organic Matter Exists Throughout the Universe". ScienceDaily. 26 October 2011.

Retrieved 27 October 2011.

Kwok, Sun; Zhang, Yong (26 October 2011). "Mixed aromatic–aliphatic organic nanoparticles as carriers of unidentified infrared emission features". Nature. 479 (7371): 80–3. Bibcode:2011Natur.479...80K. doi:10.1038/nature10542. PMID 22031328.

Then, Ker (29 August 2012). "Sugar Found in Space". National Geographic. Retrieved 31 August 2012.

"Sweet! Astronomers spot sugar molecule near star". Associated Press. 29 August 2012. Retrieved 31 August 2012.

Jorgensen, Jess K.; Favre, Cécile; Bishop, Suzanne E.; Bourke, Tyler L.; van Dishoeck, Ewine F.; Schmalzl, Markus (September 2012). "Detection of the simplest sugar, glycolaldehyde, in a solar-type protostar with ALMA" (PDF). The Astrophysical Journal Letters. 757 (1). L4. arXiv:1208.5498 Freely accessible. Bibcode:2012ApJ...757L...4J. doi:10.1088/2041-8205/757/1/L4.

Schenkel, Peter (May 2006). "SETI Requires a Skeptical Reappraisal". Skeptical Inquirer. Retrieved 28 June 2009.

Moldwin, Mark (November 2004). "Why SETI is science and UFOlogy is not". Skeptical Inquirer. Archived from the original on 13 March 2009.

"The Search for Extraterrestrial Intelligence (SETI) in the Optical Spectrum". The Columbus Optical SETI Observatory.

Whitmire, Daniel P.; Wright, David P. (April 1980). "Nuclear waste spectrum as evidence of technological extraterrestrial civilizations". Icarus. 42 (1): 149–156. Bibcode:1980Icar...42..149W. doi:10.1016/0019-1035(80)90253-5.

"Discovery of OGLE 2005-BLG-390Lb, the first cool rocky/icy

exoplanet". IAP.fr. 25 January 2006.

Then, Ker (24 April 2007). "Major Discovery: New Planet Could Harbor Water and Life". Space.com.

Schneider, Jean (10 September 2011). "Interactive Extra-solar Planets Catalog". The Extrasolar Planets Encyclopedia. Retrieved 30 January 2012.

Wall, Mike (4 April 2012). "NASA Extends Planet-Hunting Kepler Mission Through 2016". Space.com.

"NASA – Kepler". Archived from the original on 5 November 2013. Retrieved 4 November 2013.

Harrington, J. D.; Johnson, M. (4 November 2013). "NASA Kepler Results Usher in a New Era of Astronomy".

Tenenbaum, P.; Jenkins, J. M.; Seader, S.; Burke, C. J.; Christiansen, J. L.; Rowe, J. F.; Caldwell, D. A.; Clarke, B. D.; Li, J.; Quintana, E. V.; Smith, J. C.; Thompson, S. E.; Twicken, J. D.; Borucki, W. J.; Batalha, N. M.; Cote, M. T.; Haas, M. R.; Hunter, R. C.; Sanderfer, D. T.; Girouard, F. R.; Hall, J. R.; Ibrahim, K.; Klaus, T. C.; McCauliff, S. D.; Middour, C. K.; Sabale, A.; Uddin, A. K.; Wohler, B.; Barclay, T.; Still, M. (2013). "Detection of Potential Transit Signals in the First 12 Quarters of Kepler Mission Data". The Astrophysical Journal Supplement Series. 206: 5. arXiv:1212.2915 Freely accessible. Bibcode:2013ApJS..206....5T. doi:10.1088/0067-0049/206/1/5.

"My God, it's full of planets! They should have sent a poet" (Press release). Planetary Habitability Laboratory, University of Puerto Rico at Arecibo. 3 January 2012.

Santerne, A.; Díaz, R. F.; Almenara, J.-M.; Lethuillier, A.; Deleuil, M.; Moutou, C. (2013). "Astrophysical false positives in exoplanet transit surveys: Why do we need bright stars?". arXiv:1310.2133 Freely accessible [astro-ph.EP].

Cassan, A.; et al. (11 January 2012). "One or more bound planets per Milky Way star from microlensing observations". Nature. 481 (7380): 167–169. arXiv:1202.0903 Freely accessible. Bibcode:2012Natur.481..167C. doi:10.1038/nature10684. PMID 22237108.

Sanders, R. (4 November 2013). "Astronomers answer key question: How common are habitable planets?". newscenter.berkeley.edu.

Petigura, E. A.; Howard, A. W.; Marcy, G. W. (2013). "Prevalence of Earth-size planets orbiting Sun-like stars". Proceedings of the National Academy of Sciences. 110 (48): 19273–19278. arXiv:1311.6806 Freely accessible. Bibcode:2013PNAS..11019273P. doi:10.1073/pnas.1319909110. PMC 3845182 Freely accessible. PMID 24191033.

Strigari, L. E.; Barnabè, M.; Marshall, P. J.; Blandford, R. D. (2012). "Nomads of the Galaxy". Monthly Notices of the Royal Astronomical Society. 423 (2): 1856–1865. arXiv:1201.2687 Freely accessible. Bibcode:2012MNRAS.423.1856S. doi:10.1111/j.1365-2966.2012.21009.x. estimates 700 objects >10−6 solar masses (roughly the mass of Mars) per main-sequence star between 0.08 and 1 Solar mass, of which there are billions in the Milky Way.

Chang, Kenneth (24 August 2016). "One Star Over, a Planet That Might Be Another Earth". The New York Times. Retrieved 4 September 2016.

"DENIS-P J082303.1-491201 b". Caltech. Retrieved 8 March 2014.

Sahlmann, J.; Lazorenko, P. F.; Ségransan, D.; Martín, E. L.; Queloz, D.; Mayor, M.; Udry, S. (August 2013). "Astrometric orbit of a low-mass companion to an ultracool dwarf". Astronomy & Astrophysics. 556: 133. arXiv:1306.3225 Freely accessible.

Bibcode:2013A&A...556A.133S. doi:10.1051/0004-6361/201321871.

Aguilar, David A.; Pulliam, Christine (25 February 2013). "Future Evidence for Extraterrestrial Life Might Come from Dying Stars". Harvard-Smithsonian Center for Astrophysics. Release 2013-06. Retrieved 9 June 2017.

Borenstein, Seth (19 October 2015). "Hints of life on what was thought to be desolate early Earth". Excite. Yonkers, NY: Mindspark Interactive Network. Associated Press. Retrieved 20 October 2015.

Bell, Elizabeth A.; Boehnike, Patrick; Harrison, T. Mark; et al. (19 October 2015). "Potentially biogenic carbon preserved in a 4.1 billion-year-old zircon" (PDF). Proc. Natl. Acad. Sci. U.S.A. Washington, D.C.: National Academy of Sciences. 112 (47): 14518–21. Bibcode:2015PNAS..11214518B. doi:10.1073/pnas.1517557112. ISSN 1091-6490. PMC 4664351 Freely accessible. PMID 26483481. Retrieved 20 October 2015. Early edition published online before print.

"Chapter 3 — Philosophy: "Solving the Drake Equation". SETI League. December 2002. Retrieved 24 July 2015.

Burchell, M. J. (2006). "W(h)either the Drake equation?". International Journal of Astrobiology. 5 (3): 243–250. Bibcode:2006IJAsB...5..243B. doi:10.1017/S1473550406003107.

Aguirre, L. (1 July 2008). "The Drake Equation". Nova Science Now. PBS. Retrieved 7 March 2010.

Cohen, Jack; Stewart, Ian (2002). "Chapter 6: What does a Martian look like?". Evolving the Alien: The Science of Extraterrestrial Life. Hoboken, NJ: John Wiley and Sons. ISBN 0-09-187927-2.

Temming, M. (18 July 2014). "How many galaxies are there in the universe?". Sky & Telescope. Retrieved 17 December 2015.

Marcy, G.; Butler, R.; Fischer, D.; et al. (2005). "Observed Properties of Exoplanets: Masses, Orbits and Metallicities". Progress of Theoretical Physics Supplement. 158: 24–42. arXiv:astro-ph/0505003 Freely accessible. Bibcode:2005PThPS.158...24M. doi:10.1143/PTPS.158.24. Archived from the original on 2 October 2008.

Swift, Jonathan J.; Johnson, John Asher; Morton, Timothy D.; Crepp, Justin R.; Montet, Benjamin T.; et al. (January 2013). "Characterizing the Cool KOIs. IV. Kepler-32 as a Prototype for the Formation of Compact Planetary Systems throughout the Galaxy". The Astrophysical Journal. 764 (1). 105. arXiv:1301.0023 Freely accessible. Bibcode:2013ApJ...764..105S. doi:10.1088/0004-637X/764/1/105.

"100 Billion Alien Planets Fill Our Milky Way Galaxy: Study". Space.com. 2 January 2013. Archived from the original on 3 January 2013. Retrieved 10 March 2016.

"Alien Planets Revealed". Nova. Season 41. Episode 10. 8 January 2014. Event occurs at 50:56.

Overbye, Dennis (3 August 2015). "The Flip Side of Optimism About Life on Other Planets". The New York Times. Retrieved 29 October 2015.

"Who discovered that the Sun was a Star?". Stanford Solar Center.

Crowe, Michael J. (1999). The Extraterrestrial Life Debate, 1750–1900. Courier Dover Publications. ISBN 0-486-40675-X.

Wiker, Benjamin D. (4 November 2002). "Alien Ideas: Christianity and the Search for Extraterrestrial Life". Crisis Magazine. Archived from the original on 10 February 2003.

Irwin, Robert (2003). The Arabian Nights: A Companion. Tauris Parke Paperbacks. p. 204 & 209. ISBN 1-86064-983-1.

David A. Weintraub (2014). "Islam," Religions and Extraterrestrial Life (pp 161-168). Springer International Publishing.

Fontenelle, Bernard le Bovier (1990). Conversations on the Plurality of Worlds. trans. H. A. Hargreaves. University of California Press. ISBN 978-0-520-91058-4.

"Flammarion, (Nicolas) Camille (1842–1925)". The Internet Encyclopedia of Science.

"Giordano Bruno: On the Infinite Universe and Worlds (De l'Infinito Universo et Mondi) Introductory Epistle: Argument of the Third Dialogue". Archived from the original on 13 October 2014. Retrieved 4 October 2014.

"Rheita.htm". cosmovisions.com.

Evans, J. E.; Maunder, E. W. (June 1903). "Experiments as to the actuality of the "Canals" observed on Mars". Monthly Notices of the Royal Astronomical Society. 63 (8): 488–499. Bibcode:1903MNRAS..63..488E. doi:10.1093/mnras/63.8.488.

Wallace, Alfred Russel (1907). Is Mars Habitable? A Critical Examination of Professor Lowell's Book "Mars and Its Canals," With an Alternative Explanation. London: Macmillan. OCLC 8257449.

Chambers, Paul (1999). Life on Mars; The Complete Story. London: Blandford. ISBN 0-7137-2747-0.

Cross, Anne (2004). "The Flexibility of Scientific Rhetoric: A Case Study of UFO Researchers". Qualitiative Sociology. 27 (1): 3–34. doi:10.1023/B:QUAS.0000015542.28438.41.

Ailleris, Philippe (January–February 2011). "The lure of local SETI: Fifty years of field experiments". Acta Astronautica. 68 (1–2): 2–15. Bibcode:2011AcAau..68....2A. doi:10.1016/j.actaastro.2009.12.011.

"LECTURE 4: MODERN THOUGHTS ON EXTRATERRESTRIAL LIFE". The University of Antarctica. Retrieved 25 July 2015.

Ward, Peter; Brownlee, Donald (2000). Rare Earth: Why Complex Life is Uncommon in the Universe. Copernicus. Bibcode:2000rewc.book.....W. ISBN 978-0-387-98701-9.

"Hawking warns over alien beings". BBC News. 25 April 2010. Retrieved 2 May 2010.

Diamond, Jared M. (2006). "Chapter 12". The Third Chimpanzee: The Evolution and Future of the Human Animal. Harper Perennial. ISBN 978-0-06-084550-6.

Larson, Phil (5 November 2011). "Searching for ET, But No Evidence Yet". White House. Archived from the original on 24 November 2011. Retrieved 6 November 2011.

Atkinson, Nancy (5 November 2011). "No Alien Visits or UFO Coverups, White House Says". UniverseToday. Retrieved 6 November 2011.

186.Jump up ^ "Special Issue: Exoplanets". Science. 3 May 2013. Retrieved 18 May 2013.

Chang, Kenneth (17 April 2014). "Scientists Find an 'Earth Twin', or Maybe a Cousin". The New York Times.

Borenstein, Seth (13 February 2015). "Should We Call the Cosmos Seeking ET? Or Is That Risky?". The New York Times. Associated Press. Archived from the original on 14 February 2015.

Ghosh, Pallab (12 February 2015). "Scientist: 'Try to contact aliens'". BBC News. Retrieved 12 February 2015.

"Regarding Messaging to Extraterrestrial Intelligence (METI) / Active Searches for Extraterrestrial Intelligence (Active SETI)". University of California, Berkeley. 13 February 2015. Retrieved 14

February 2015.

Katz, Gregory (20 July 2015). "Searching for ET: Hawking to look for extraterrestrial life". Excite!. Associated Press. Retrieved 20 July 2015.

"The Drake Equation", SETI, Retrieved 14 April 2018

Hannah Osborne (On 7/26/17 at 4:45 AM), "The Fermi Paradox: Why Haven't We Found Alien Life Anywhere Else in the Universe?", Retrieved 14 April 2018.

Mindy Weisberger, Senior Writer (May 18, 2017 07:15am ET) , "How Do Scientists Search for Extraterrestrial Life?" Retrieved 15 April 2018.

Robert Trundle, PhD, "Is ET Here?" ISBN 0-9735341-2-5, Retrieved 15 April 2018.

Andrew Whalen (On 3/14/18 at 12:31 PM), "Stephen Hawking on Alien Life, Extraterrestrials and the Possibility of UFOs Visiting Earth", Retrieved 16 April 2018.

Guy Birchall, (13th December 2017, 1:15 pm), OUT OF THIS WORLD

"Why Stephen Hawking and the world's top scientists say this massive object hurtling through space could be an alien spaceship", Retrieved 15 April 2018.

Bridget Brennan, wires, (Updated 21 Jul 2015, 2:23amTue 21 Jul 2015, 2:23am), Retrieved 16 April 2018.

Marianne Spoon, "How Carl Sagan Worked", Retrieved 16 April 2018.

Kim LaCapria, "aliens disclose.tv NASA +2 more NASA

conspiracies Waterford whispers",

(19 September 2016), Retrieved 15 April 2018.

June 16, 2017 (June 16, 2017) " Analysis & Implications, Analysis - General, Evidence & Phenomenon, FAQ, General / Overview, Government Cover-Up, Government Cover-Up - General, Government UFO Documents, Hints & Tips, Latest News, Mainstream Science & UFOs, Media & UFOs, Sighting Reports", Retrieved 15 April 2018.

UFO International Project » Analysis & Implications » Scientists Confirm That Newly discovered STAR system, could be home to interplanetary ALIENS (Drip Feeding Disclosure), March 18, 2017 (March 18, 2017) "Analysis & Implications, Analysis - General, Evidence & Phenomenon, General / Overview, Government Cover-Up, Government Cover-Up - General, Government UFO Documents, Latest News, Mainstream Science & UFOs, Media & UFOs", Retrieved 16 April 2018.

March 9, 2017 (March 9, 2017) " Alien beings sightings, Alien/UFO contact techniques, All other, All other, Analysis & Implications, Analysis - General, CIA, Evidence & Phenomenon, General / Overview, General/Mass Sightings, GEPAN / SEPRA (France), Government Cover-Up, Government Cover-Up - General, Government Studies, Government UFO Documents, Govt. & Scientific Studies, International Sightings, Interstellar Travel, KGB, Latest News, Life in the Universe, Life in the Universe, Mainstream Science & UFOs, Majestic Documents, Media & UFOs, MI5/MI6, NASA, NSA, Overview / General, Philosophy of Science, Religion and UFOs, SETI (Search for Extra-terrestrial Intelligence), Skeptics & Their Arguments, Speed of Light Limit, UFO Topics, UFOS in the Rest Of The World", Retrieved 16 April 2018.

Burchell, M.J. (2006). "W(h)either the Drake equation?". International Journal of Astrobiology. 5 (3): 243–250. Bibcode:2006IJAsB...5..243B. doi:10.1017/S1473550406003107.

Glade, N.; Ballet, P.; Bastien, O. (2012). "A stochastic process approach of the drake equation parameters". International Journal of Astrobiology. 11 (2): 103–108. arXiv:1112.1506 Freely accessible. Bibcode:2012IJAsB..11..103G. doi:10.1017/S1473550411000413.

"Chapter 3 — Philosophy: "Solving the Drake Equation". Ask Dr. SETI. SETI League. December 2002. Retrieved 2013-04-10.

Drake, N. (30 June 2014). "How my Dad's Equation Sparked the Search for Extraterrestrial Intelligence". National Geographic. Retrieved 2 October 2016.

Cocconi, G.; Morisson, P. (1959). "Searching for Interstellar Communications" (PDF). Nature. 184 (4690): 844–846. Bibcode:1959Natur.184..844C. doi:10.1038/184844a0. Retrieved 2013-04-10.

Schilling, G.; MacRobert, A. M. (2013). "The Chance of Finding Aliens". Sky & Telescope. Retrieved 2013-04-10.

newspaper, staff (8 November 1959). "Life on Other Planets?". Sydney Morning Herald. Retrieved 2015-10-02.

"The Drake Equation Revisited: Part I". Astrobiology Magazine. Sep 29, 2003. Retrieved May 20, 2017.

Zaun, H. (1 November 2011). "Es war wie eine 180-Grad-Wende von diesem peinlichen Geheimnis!" [It was like a 180 degree turn from this embarrassing secret]. Telepolis (in German). Retrieved 2013-08-13.

"Drake Equation Plaque". Retrieved 2013-08-13.

Darling, D. J. "Green Bank conference (1961)". The Encyclopedia of Science. Retrieved 2013-08-13.

Aguirre, L. (1 July 2008). "The Drake Equation". Nova Science Now. PBS. Retrieved 2010-03-07.

"What do we need to know about to discover life in space?". SETI Institute. Retrieved 2013-04-16.

Jones, D. S. (26 September 2001). "Beyond the Drake Equation". Retrieved 2013-04-17.

"The Search for Life : The Drake Equation 2010 - Part 1". BBC Four. 2010. Retrieved 2013-04-17.

SETI: A celebration of the first 50 years. Keith Cooper. Astronomy Now. 2000

Hetesi, Z.; Regaly, Z. (2006). "A new interpretation of Drake-equation" (PDF). Journal of the British Interplanetary Society. 59: 11–14. Bibcode:2006JBIS...59...11H.[permanent dead link]

Maccone, C. (2010). "The Statistical Drake Equation". Acta Astronautica. 67 (11–12): 1366–1383. Bibcode:2010AcAau..67.1366M. doi:10.1016/j.actaastro.2010.05.003.

Brin, G. D. (1983). "The Great Silence – The Controversy Concerning Extraterrestrial Intelligent Life". Quarterly Journal of the Royal Astronomical Society. 24 (3): 283–309. Bibcode:1983QJRAS..24..283B.

Zaitsev, A. (May 2005). "The Drake Equation: Adding a METI Factor". SETI League. Retrieved 2013-04-20.

Jones, Chris (December 7, 2016). "'The World Sees Me as the One Who Will Find Another Earth' - The star-crossed life of Sara Seager, an astrophysicist obsessed with discovering distant planets". New York Times. Retrieved December 8, 2016.

The Drake Equation Revisited: Interview with Planet Hunter Sara Seager Devin Powell, Astrobiology Magazine 4 September 2013.

"A New Equation Reveals Our Exact Odds of Finding Alien Life". io9.

Drake, F.; Sobel, D. (1992). Is Anyone Out There? The Scientific Search for Extraterrestrial Intelligence. Delta. pp. 55–62. ISBN 0-385-31122-2.

Glade, N.; Ballet, P.; Bastien, O. (2012). "A stochastic process approach of the drake equation parameters". International Journal of Astrobiology. 11 (2): 103–108. arXiv:1112.1506 Freely accessible. Bibcode:2012IJAsB..11..103G. doi:10.1017/S1473550411000413. Note: This reference has a table of 1961 values, claimed to be taken from Drake & Sobel, but these differ from the book.

Robitaille, Thomas P.; Barbara A. Whitney (2010). "The present-day star formation rate of the Milky Way determined from Spitzer-detected young stellar objects". The Astrophysical Journal Letters. 710 (1): L11. arXiv:1001.3672 Freely accessible. Bibcode:2010ApJ...710L..11R. doi:10.1088/2041-8205/710/1/L1.

Wanjek, C. (2 Jul 2015). "The Drake Equation". Cambridge University Press. Retrieved 2016-09-09.

Kennicutt, Robert C.; Evans, Neal J. (22 September 2012). "Star Formation in the Milky Way and Nearby Galaxies". Annual Review of Astronomy and Astrophysics. 50 (1): 531–608. arXiv:1204.3552 Freely accessible. Bibcode:2012ARA&A..50..531K. doi:10.1146/annurev-astro-081811-125610.

Palmer, J. (11 January 2012). "Exoplanets are around every star, study suggests". BBC. Retrieved 2012-01-12.

Cassan, A.; et al. (11 January 2012). "One or more bound planets per Milky Way star from microlensing observations". Nature. 481 (7380): 167–169. arXiv:1202.0903 Freely accessible. Bibcode:2012Natur.481..167C. doi:10.1038/nature10684. PMID 22237108.

Overbye, Dennis (4 November 2013). "Far-Off Planets Like the Earth Dot the Galaxy". New York Times. Retrieved 5 November 2013.

Petigura, Eric A.; Howard, Andrew W.; Marcy, Geoffrey W. (31 October 2013). "Prevalence of Earth-size planets orbiting Sun-like stars". Proceedings of the National Academy of Sciences of the United States of America. 110: 19273–19278. arXiv:1311.6806 Freely accessible. Bibcode:2013PNAS..11019273P. doi:10.1073/pnas.1319909110. PMC 3845182 Freely accessible. PMID 24191033.

Khan, Amina (4 November 2013). "Milky Way may host billions of Earth-size planets". Los Angeles Times. Retrieved 5 November 2013.

Govert Schilling (November 2011). "The Chance of Finding Aliens: Reevaluating the Drake Equation". astro-tom.com.

Trimble, V. (1997). "Origin of the biologically important elements". Origins of Life and Evolution of the Biosphere. 27 (1–3): 3–21. doi:10.1023/A:1006561811750. PMID 9150565.

Lineweaver, C. H.; Fenner, Y.; Gibson, B. K. (2004). "The Galactic Habitable Zone and the Age Distribution of Complex Life in the Milky Way". Science. 303 (5654): 59–62. arXiv:astro-ph/0401024 Freely accessible. Bibcode:2004Sci...303...59L. doi:10.1126/science.1092322. PMID 14704421.

Dressing, C. D.; Charbonneau, D. (2013). "The Occurrence Rate of Small Planets around Small Stars". The Astrophysical Journal. 767: 95. arXiv:1302.1647 Freely accessible.

Bibcode:2013ApJ...767...95D. doi:10.1088/0004-637X/767/1/95.

"Red Dwarf Stars Could Leave Habitable Earth-Like Planets Vulnerable to Radiation". SciTech Daily. Retrieved 22 September 2015.

Heller, René; Barnes, Rory (29 April 2014). "Constraints on the Habitability of Extrasolar Moons". Proceedings of the International Astronomical Union. 8 (S293): 159–164. arXiv:1210.5172 Freely accessible. Bibcode:2014IAUS..293..159H. doi:10.1017/S1743921313012738.

Ward, Peter D.; Brownlee, Donald (2000). Rare Earth: Why Complex Life is Uncommon in the Universe. Copernicus Books (Springer Verlag). ISBN 0-387-98701-0.

Davies, P. (2007). "Are Aliens Among Us?". Scientific American. 297 (6): 62–69. Bibcode:2007SciAm.297f..62D. doi:10.1038/scientificamerican1207-62.

Crick, F. H. C.; Orgel, L. E. (1973). "Directed Panspermia" (PDF). Icarus. 19 (3): 341–346. Bibcode:1973Icar...19..341C. doi:10.1016/0019-1035(73)90110-3.

43.^ Jump up to: a b "Ernst Mayr on SETI". The Planetary Society. Archived from the original on 6 December 2010.

Rare Earth, p. xviii.: "We believe that life in the form of microbes or their equivalents is very common in the universe, perhaps more common than even Drake or Sagan envisioned. However, complex life – animals and higher plants – is likely to be far rarer than commonly assumed."

Campbell, A. (13 March 2005). "Review of Life's Solution by Simon Conway Morris". Archived from the original on 16 July 2011.

Bonner, J. T. (1988). The evolution of complexity by means of natural selection. Princeton University Press. ISBN 0-691-08494-7.

Forgan, D.; Elvis, M. (2011). "Extrasolar Asteroid Mining as Forensic Evidence for Extraterrestrial Intelligence". International Journal of Astrobiology. 10 (4): 307–313. arXiv:1103.5369 Freely accessible. Bibcode:2011IJAsB..10..307F. doi:10.1017/S1473550411000127.

J. Tarter (September 2001). "The Search for Extraterrestrial Intelligence (SETI)". Annual Review of Astronomy and Astrophysics. 39: 511–548. Bibcode:2001ARA&A..39..511T. doi:10.1146/annurev.astro.39.1.511.

Shermer, M. (August 2002). "Why ET Hasn't Called". Scientific American: 21.

Grinspoon, D. (2004). Lonely Planets.

Goldsmith, D.; Owen, T. (1992). The Search for Life in the Universe (2nd ed.). Addison-Wesley. p. 415. ISBN 1-891389-16-5.

"The value of N remains highly uncertain. Even if we had a perfect knowledge of the first two terms in the equation, there are still five remaining terms, each of which could be uncertain by factors of 1,000." from Wilson, TL (2001). "The search for extraterrestrial intelligence". Nature. Nature Publishing Group. 409 (6823): 1110–1114. Bibcode:2001Natur.409.1110W. doi:10.1038/35059235., or more informally, "The Drake Equation can have any value from "billions and billions" to zero", Michael Crichton, as quoted in Douglas A. Vakoch; et al. (2015). The Drake Equation: Estimating the prevalence of extraterrestrial life through the ages. Cambridge University Press. ISBN 978-1-10-707365-4., pp. 13

"The Drake Equation". psu.edu.

Devin Powell, Astrobiology Magazine. "The Drake Equation Revisited: Interview with Planet Hunter Sara Seager". Space.com.

Govert Schilling; Alan M. MacRobert (June 3, 2009). "The Chance of Finding Aliens". Sky & Telescope.

Dean, T. (10 August 2009). "A review of the Drake Equation". Cosmos Magazine. Archived from the original on 3 June 2013. Retrieved 16 April 2013.

Rare Earth, page 270: "When we consider factors such as the abundance of planets and the location and lifetime of the habitable zone, the Drake Equation suggests that only between 1% and 0.001% of all stars might have planets with habitats like Earth. [...] If microbial life forms readily, then millions to hundreds of millions of planets in the galaxy have the potential for developing advanced life. (We expect that a much higher number will have microbial life.)"

von Bloh, W.; Bounama, C.; Cuntz, M.; Franck, S. (2007). "The habitability of super-Earths in Gliese 581". Astronomy & Astrophysics. 476 (3): 1365–1371. arXiv:0705.3758 Freely accessible. Bibcode:2007A&A...476.1365V. doi:10.1051/0004-6361:20077939.

Selsis, F.; Kasting, J. F.; Levrard, B.; Paillet, J.; Ribas, I.; Delfosse, X. (2007). "Habitable planets around the star Gliese 581?". Astronomy & Astrophysics. 476 (3): 1373–1387. arXiv:0710.5294 Freely accessible. Bibcode:2007A&A...476.1373S. doi:10.1051/0004-6361:20078091.

Lineweaver, C. H.; Davis, T. M. (2002). "Does the rapid appearance of life on Earth suggest that life is common in the universe?". Astrobiology. 2 (3): 293–304. arXiv:astro-ph/0205014 Freely accessible. Bibcode:2002AsBio...2..293L. doi:10.1089/153110702762027871. PMID 12530239.

Forgan, D. (2009). "A numerical testbed for hypotheses of extraterrestrial life and intelligence". International Journal of

Astrobiology. 8 (2): 121–131. arXiv:0810.2222 Freely accessible. Bibcode:2009IJAsB...8..121F. doi:10.1017/S1473550408004321.

"Are we alone? Setting some limits to our uniqueness". phys.org. 28 April 2016.

"Are We Alone? Galactic Civilization Challenge". PBS Space Time. 5 October 2016. PBS Digital Studios.

Ellie Zolfagharifard (28 April 2016). "An advanced alien civilization DID exist before us: Scientists update prediction of alien life using latest Kepler exoplanet data". Daily Mail (London).

Adam Frank (10 June 2016). "Yes, There Have Been Aliens". New York Times.

A. Frank; W.T. Sullivan III; (22 April 2016). "A New Empirical Constraint on the Prevalence of Technological Species in the Universe". Astrobiology (published 13 May 2016). 16 (5): 359–362. arXiv:1510.08837 Freely accessible. Bibcode:2016AsBio..16..359F. doi:10.1089/ast.2015.1418. PMID 27105054.

Dvorsky, G. (31 May 2007). "The Drake Equation is obsolete". Sentient Developments. Retrieved 2013-08-21.

Tarter, J. (May–June 2006). "The Cosmic Haystack Is Large". Skeptical Inquirer. 30 (3). Retrieved 2013-08-21.

Alexander, A. "The Search for Extraterrestrial Intelligence: A Short History - Part 7: The Birth of the Drake Equation". The Planetary Society. Archived from the original on 2005-03-06.

Christopher J. Conselice; et al. (2016). "The Evolution of Galaxy Number Density at z < 8 and its Implications". The Astrophysical Journal. 830 (2): 83. arXiv:1607.03909v2 Freely accessible. Bibcode:2016ApJ...830...83C. doi:10.3847/0004-637X/830/2/83.

Fountain, Henry (17 October 2016). "Two Trillion Galaxies, at the

Very Least". New York Times. Retrieved 17 October 2016.

Jones, E. M. (1 March 1985). "Where is everybody?" An account of Fermi's question (PDF) (Report). Los Alamos National Laboratory. Bibcode:1985STIN...8530988J. doi:10.2172/5746675 Freely accessible. OSTI 5746675 Freely accessible. Retrieved 2013-08-21.

Krauthammer, C. (29 December 2011). "Are we alone in the Universe?". The Washington Post. Retrieved 2013-08-21.

Webb, S. (2015). If the Universe Is Teeming with Aliens ... WHERE IS EVERYBODY?: Seventy-Five Solutions to the Fermi Paradox and the Problem of Extraterrestrial Life. Springer International Publishing. ISBN 3319132350.

Hanson, R. (15 September 1998). "The Great Filter — Are We Almost Past It?". Retrieved 2013-08-21.

The Making of Star Trek by Stephen E. Whitfield and Gene Roddenberry, Ballantine Books, N. Y.,

INDEX

"

"The Order of the Dolphin", 15

1

19th century, 38

2

20th Century, 39

A

Active SETI, 73
alien life, 23
America, **iii**
Arecibo Observatory in Puerto Rico, 57
Artificial Martian channels, depicted by Percival Lowell, 39

B

Bibliography, 166
Brad Gibson, Yeshe Fenner, and Charley, 17
Breakthrough Listen, 61

C

Carl Sagon, 141
Ceres, 29
Cocconi and Morrison published their article, 14
Community SETI projects, 62
Cosmic Pluralism, 37
Current Estimates, 16

D

Different background model, 111
Direct search, 32

E

Early Modern Period, 37
Early work, 51
Enceladus, 30
Energy Policy, **10**
Europa, 29
European Space Agency, 16
Evolution, 25
Extrasolar Planets, 35
Extraterrestrial Life, 23

F

Frank Drake, 1961, 12

G

Gamma-ray Bursts, 67

H

Harvard University astronomy professor Harlow Shapley, 14
History, 51
How Do Scientists Search for Extraterrestrials?, 76

I

Indirect search, 34

J

John Grunsfeld, a physics PhD, 11

Jupiter, 29
Jupiter system, 29

K

Kepler spacecraft, 11
Kevin Hand, 11

M

Mars, 28
Matt Mountain, 10
Milky Way Galaxy, 16

N

NASA, 10, 149
NASA About To Confirm They Have Discovered Alien Life? Or Just More Lies And Waffle?, 152

O

Official Alien Contact And Humanity's Path To The Universe, 160
Ongoing radio searches, 58
Optical Experiments, 65
Other bodies, 31

P

Planetary Habitability in the Solar System, 26
Post Detection Disclosure Protocol, 71

R

Recent history, 40

S

Sara Seager, 10

Saturn System, 30
Scientific search, 32
Search for extraterrestrial artifacts, 68
Search for Extraterrestrial Intelligence, 51
Stephen Hawking, 120
Stephen Hawking on Alien Life, Extraterrestrials and the Possibility of UFOs Visiting Earth, 121

T

Technosignatures, 70
Terrestrial Analysis, 36
The Drake Equation, 12
The Fermi Paradox, 44
The *Kepler* space telescope, 35
The SETI League and Project Argus, 64
The WOW! Signal, 53
Titan, 31

U

Using Artificial Intelligence to Search for Extraterrestrial Intelligence, 90

V

Venus, 28

W

Wernher Von Braun, 114
Who is Russian billionaire Yuri Milner?, 133
Why Stephen Hawking and the world's top scientists say this massive object hurtling through space could be an alien spaceship, 123

www.ingramcontent.com/pod-product-compliance
Lightning Source LLC
Chambersburg PA
CBHW031926240526
45464CB00023B/1712